Membrane Distillation

Special Issue Editor

Enrico Drioli

MDPI • Basel • Beijing • Wuhan • Barcelona • Belgrade

MDPI

Special Issue Editor
Enrico Drioli
University of Calabria
Italy

Editorial Office
MDPI AG
St. Alban-Anlage 66
Basel, Switzerland

This edition is a reprint of the Special Issue published online in the open access journal *Applied Sciences* (ISSN 2076-3417) in 2017 (available at: http://www.mdpi.com/journal/applsci/special_issues/Membrane_Distillation).

For citation purposes, cite each article independently as indicated on the article page online and as indicated below:

Author 1; Author 2. Article title. *Journal Name* **Year**, *Article number*, page range.

First Edition 2017

ISBN 978-3-03842-460-4 (Pbk)
ISBN 978-3-03842-461-1 (PDF)

Photo courtesy of Enrico Drioli

Table of Contents

About the Special Issue Editor..v

Enrico Drioli, Aamer Ali and Francesca Macedonio
Membrane Operations for Process Intensification in Desalination
Reprinted from: *Appl. Sci.* **2017**, 7(1), 100; doi: 10.3390/app7010100 1

Minwei Yao, Yun Chul Woo, Leonard D. Tijing, Cecilia Cesarini and Ho Kyong Shon
Improving Nanofiber Membrane Characteristics and Membrane Distillation Performance
of Heat-Pressed Membranes via Annealing Post-Treatment
Reprinted from: *Appl. Sci.* **2017**, 7(1), 78; doi: 10.3390/app7010078 15

Faisal Abdulla AlMarzooqi, Muhammad Roil Bilad and Hassan Ali Arafat
Improving Liquid Entry Pressure of Polyvinylidene Fluoride (PVDF) Membranes by Exploiting
the Role of Fabrication Parameters in Vapor-Induced Phase Separation VIPS and Non-Solvent-
Induced Phase Separation (NIPS) Processes
Reprinted from: *Appl. Sci.* **2017**, 7(2), 181; doi: 10.3390/app7020181 26

**Chia-Chieh Ko, Chien-Hua Chen, Yi-Rui Chen, Yu-Hsun Wu, Soon-Chien Lu, Fa-Chun Hu,
Chia-Ling Li and Kuo-Lun Tung**
Increasing the Performance of Vacuum Membrane Distillation Using Micro-Structured
Hydrophobic Aluminum Hollow Fiber Membranes
Reprinted from: *Appl. Sci.* **2017**, 7(4), 357; doi: 10.3390/app7040357 41

**Lies Eykens, Kristien De Sitter, Chris Dotremont, Wim De Schepper, Luc Pinoy and
Bart Van Der Bruggen**
Wetting Resistance of Commercial Membrane Distillation Membranes in Waste Streams
Containing Surfactants and Oil
Reprinted from: *Appl. Sci.* **2017**, 7(2), 118; doi: 10.3390/app7020118 51

Daniel Woldemariam, Andrew Martin and Massimo Santarelli
Exergy Analysis of Air-Gap Membrane Distillation Systems for Water Purification Applications
Reprinted from: *Appl. Sci.* **2017**, 7(3), 301; doi: 10.3390/app7030301 63

**Mourad Laqbaqbi, Julio Antonio Sanmartino, Mohamed Khayet, Carmen García-Payo and
Mehdi Chaouch**
Fouling in Membrane Distillation, Osmotic Distillation and Osmotic Membrane Distillation
Reprinted from: *Appl. Sci.* **2017**, 7(4), 334; doi: 10.3390/app7040334 74

Joanna Kujawa, Sophie Cerneaux, Wojciech Kujawski and Katarzyna Knozowska
Hydrophobic Ceramic Membranes for Water Desalination
Reprinted from: *Appl. Sci.* **2017**, 7(4), 402; doi: 10.3390/app7040402 114

About the Special Issue Editor

Enrico Drioli is Emeritus Professor at the School of Engineering of the University of Calabria and Founding Director of the Institute on Membrane Technology, CNR, Italy. Since 2012, he has been Distinguished Adjunct Professor at CEDT King Abdulaziz University, Jeddah, Saudi Arabia; since 2010, he has been WCU Distinguish Visiting Professor at the Hanyang University, Seoul, Korea. His research activities focus on Membrane Science and Engineering, Membranes in Artificial Organs, Integrated Membrane Processes, Membrane Preparation and Transport Phenomena in Membranes, Membrane Distillation and Membrane Contactors, and Catalytic Membrane and Catalytic Membrane Reactors. He is involved in many International Societies, Scientific Committees, Editorial Boards, and International Advisory Boards. Currently, he is Chairman of the Section on "Membrane Engineering" of the European Federation of Chemical Engineering and coordinator of EU-EUDIME Doctorate School on Membrane Engineering. He has been coordinator of several international research projects. He is Honorary President of European Membrane Society (1999).

He is the recipient of various Awards and Honours, ex. "Richard Maling Barrer Prize" of the EMS, Academician Semenov Medal of Russian Academy of Engineering Science, MIAC International Award for his contributions in the field of Membrane Science and Technologies, etc. He is author of more than 800 scientific papers, 22 patents and 24 books on Membrane Science and Technology.

applied
sciences

MDPI

Article

Membrane Operations for Process Intensification in Desalination

Enrico Drioli [1,2,3,4], Aamer Ali [1,]* and Francesca Macedonio [1,2,]*

1   Institute on Membrane Technology (ITM-CNR), National Research Council, c/o The University of Calabria, Cubo 17C, Via Pietro Bucci, Rende 87036, Italy; e.drioli@itm.cnr.it
2   Department of Environmental and Chemical Engineering, University of Calabria, Rende 87036, Italy
3   WCU Energy Engineering Department, Hanyang University, Seoul 133-791, Korea
4   Center of Excellence in Desalination Technology, King Abdulaziz University, Jeddah 21589, Saudi Arabia
*   Correspondence: amir_hmmad@hotmail.com (A.A.); francesca.macedonio@unical.it or f.macedonio@itm.cnr.it (F.M.); Tel.: +39-0984-492014 (A.A.); +39-0984-492012 (F.M.)

Academic Editor: Mohamed Khayet
Received: 16 December 2016; Accepted: 13 January 2017; Published: 20 January 2017

Abstract: Process intensification strategy (PIS) is emerging as an interesting guideline to revolutionize process industry in terms of improved efficiency and sustainability. Membrane engineering has appeared as a strong candidate to implement PIS. The most significant progress has been observed in desalination where substantial reduction in overall energy demand, environmental footprint, and process hazards has already been accomplished. Recent developments in membrane engineering are shaping the desalination industry into raw materials and energy production where fresh water will be produced as a byproduct. The present study discusses the current and perspective role of membrane engineering in achieving the objectives of PIS in the field of desalination.

Keywords: process engineering; membrane operations; desalination; metrics

1. Introduction

During the last 50 years, the world's population has doubled and gross domestic production has increased ten folds, reflecting the underlying massive industrialization during this period. These developments have put the resources of freshwater, energy, and raw materials under ever-growing strain. Energy consumption has increased by five times during the last five decades and the majority of this energy consumption is coming from finite and polluting fossil fuels [1]. In the water sector, it has been estimated that two thirds of the world's population might be facing insufficient access to clean freshwater by 2025 [2]. Similarly, traditional mining is facing several environmental and sustainability related concerns. The scenario places an emphasis on sustainable industrial growth across the globe that can be realized by using the material and energy resources more efficiently and by exploiting the nontraditional but sustainable resources of these commodities while, at the same time, eliminating or minimizing the environmental hazards associated with the related processes [3]. These requirements clearly point out the urgency to develop new processes capable of producing and using energy, freshwater, and raw materials more efficiently and with the potential to exploit alternative resources of these products.

Lack of a precise definition of sustainable development has resulted in the evolution of specific guidelines such as the PIS to implement the concept of sustainable development. Process intensification as defined by Stankiewicz and Moulijn [3,4] is the development of novel equipment and techniques that, compared to those commonly used, dramatically improve manufacturing and processing by decreasing substantially equipment size, improving raw material to production ratio, decreasing energy consumption and waste production, and that ultimately results in cheaper, efficient, safer, and more sustainable technologies. Modern membrane engineering represents one of the most interesting ways for developing processes in accordance with the guidelines provided by PIS to meet the challenges of the modern world [5,6]. The main features of membrane engineering which make it perfectly aligned with PIS include its high selectivity and permeability for transport of specific components, the ease with which it can integrate with other processes or other membrane operations, its tendency to be less energy intensive and highly efficient, as well as its tendency to have low capital costs, small footprints, and high safety, operational simplicity, and flexibility [5,7–9]. These exceptional features extend the sphere of applications of membrane engineering from water, energy, and raw materials sectors to sophisticated biomedical applications.

Desalination represents one of the industrial sectors where membrane engineering has emerged as the key player to successfully implement the concept of PIS. In many parts of the world, conventional thermal desalination plants have been replaced with far less energy intensive, compact, and safer reverse osmosis (RO) units. Currently RO occupies more than a 60% share of the desalination market [10]. Thermal processes are limited only to the regions with abundant sources of fossil fuels, such as the Middle East. However, despite their loudly spoken advantages, the pressure driven processes face some challenges that hinder their wide spread and uniform growth in water stressed regions across the globe: operation at high pressure, high energy consumption, utilization of high grade energy, limited recovery factor (typically 40%–50%), and disposal of brine are the most significant obstacles that negatively affect the process economy and cause environmental problems. Several investigations have been carried out to tackle these challenges. These investigations have focused on improving the process performance of RO by optimizing the operating conditions and process design, by introducing a better control system, and by integrating the process with renewable energy resources. In order to optimize the process, the impact of energy recovery devices, membrane permeability, process configuration, brine management cost, pump efficiency, and frictional pressure drop on specific water cost under the constraints imposed by the osmotic pressure has been analyzed [11]. In another study, it was concluded that further reduction in specific water cost in RO is less likely to take place through the development of more permeable membranes, rather it should come from better fouling and scaling control, improved brine management, lower pretreatment cost, improved process control, and process optimization [12]. Improved control systems have been developed to incorporate the variation in feed water salinity, large set-point changes, and for the optimum management and operation of integrated wind-solar energy generation and RO desalination systems [12–14].

In addition to the abovementioned approaches, the use of new membrane operations offer promising solutions to these challenges. The use of the new processes in combination with the traditional ones can not only resolve the problem of waste handling but can also provide the opportunity to boost the economy of the process. An integrated approach takes into account energy savings (also production in certain cases), water rationalization, minimization of chemical utilization, resource recovery, and waste production [4]. Therefore, integrated systems can contribute significantly to the solution of strategic aspects of industrial productions. Different membrane operations can be coupled in integrated systems in order to approach the ambitious objective of "zero liquid discharge". A qualitative comparative analysis of various generations of desalination technology on the basis of different process intensification parameters has been explained in Figure 1. The first generation desalination plants (thermal) are bulky, consume huge amounts of energy, and have large environmental impacts in terms of carbon emissions and the use of various chemicals (e.g., anti-sealants). Due to high temperatures, the safety level of these plants is relatively weak. With the advent of RO technology, the situation has improved. However, energy consumption and environmental impact (in terms of carbon emission and brine disposal) of RO is still high. These drawbacks can be addressed by integrating novel membrane processes including pressure retarded osmosis (PRO), membrane distillation (MD), forward osmosis (FO), and reverse electrodialysis (RED). The details of these new processes have been explained in later sections.

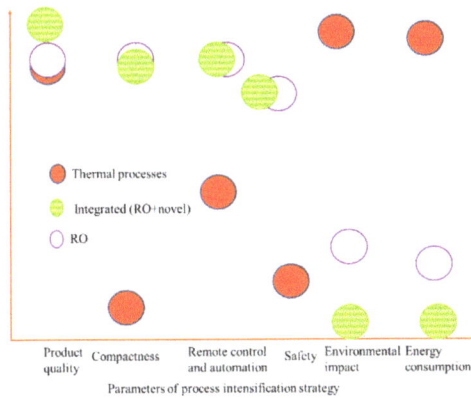

Figure 1. Qualitative analysis of conventional thermal, reverse osmosis (RO), and integrated membrane processes for desalination based on various parameters of PIS (process intensification strategy). *y*-axis indicates the qualitative comparison of the parameters list along *x*-axis for different processes (thermal, integrated, RO alone).

New Metrics for PIS (Process Intensification Strategy)

For an easy and fair comparison of conventional unit operations with membrane technology, Criscuoli and Drioli [15] have proposed some new metrics. The metrics allow for the monitoring of the progress and improvement of membrane operations in the logic of process intensification taking into account plant size, weight, flexibility, modularity, etc. Overall assessment of sustainable processes should also consider existing metrics (mass and waste intensity), environmental factors, economic, and society indicators. Mass and waste intensities (Equations (1) and (2), respectively) are used to quantify the amount of product which is produced from a particular process with respect to the amount of input materials or waste produced from the process. Reduction in mass and waste intensities is the preferred form of improvement of the process. Membrane operations have small footprints and, therefore, can be used to quantify how productivity is influenced by plant size or weight (Equations (3) and (4), respectively). The productivity to weight ratio is of particular interest if the plant is constructed off-shore or in remote areas. Taking into account the entire lifetime of the plant, it is important that it has flexibility and modularity (quantified by Equations (5) and (6), respectively), so it can be adjusted according to changes in the productivity, variation in pressure, temperature, feed compositions, or other process related parameters. The modularity equation considers changes (increase/decrease) in plant size compared to the productivity. These metrics should also be higher than 1 for a membrane plant to be preferred [15].

$$\text{Mass Intensity} = \frac{\text{Total mass [kg]}}{\text{Mass of product [kg]}} \tag{1}$$

$$\text{Waste Intensity} = \frac{\text{Total waste [kg]}}{\text{Mass of product [kg]}} \tag{2}$$

$$\text{Productivity/Size ratio} = \frac{\frac{\text{Productivity}}{\text{Size}} (\text{Membranes})}{\frac{\text{Productivity}}{\text{Size}} (\text{Tradional process})} \tag{3}$$

$$\text{Productivity/Weight ratio} = \frac{\frac{\text{Productivity}}{\text{Weight}}(\text{Membranes})}{\frac{\text{Productivity}}{\text{Weight}}(\text{Tradional process})} \quad (4)$$

$$\text{Flexibility} = \frac{\text{Variations}_{\text{handled}}(\text{Membranes})}{\text{Variations}_{\text{handled}}(\text{Tradional process})} \quad (5)$$

$$\text{Modularity} = \frac{\left|\frac{\text{area}_2}{\text{area}_1} - \frac{\text{productivity}_2}{\text{productivity}_1}\right|(\text{Membranes})}{\left|\frac{\text{area}_2}{\text{area}_1} - \frac{\text{productivity}_2}{\text{productivity}_1}\right|(\text{Tradional process})} \quad (6)$$

2. New Membrane Operations Aligned with PIS

2.1. Membrane Distillation/Crystallization

Besides the widely accepted RO in desalination which has already outclassed traditional thermal processes, new and relatively less explored membrane operations with more promises are emerging. In the desalination industry, membrane distillation and membrane crystallization, in particular, can add a positive effect on the process by increasing the overall water production and recovering valuable salts from the brine, thus approaching zero-liquid-discharge and the goals of process intensification strategy. The recovery of minerals from brine through the use of membrane distillation/crystallization might have a significant role in determining the future outlook of the mining industry. While the mining industry is under extreme stress due to excessive exploitation of minerals, the volume of brine (and therefore minerals contained in it) is increasing across the globe due to increases in the net volume desalination capacity. Seawater brine contains a number of essential as well strategic elements including sodium, magnesium, barium, lithium, etc. and can thus serve as a form of open sky mine for the recovery of these components. Dramatic increases in the consumption of some of these materials have been observed recently. For example, the use of strontium has increased in the oil and gas industry as weighing material in mud. Similarly, traditional sources of lithium might not be enough in the near future to fulfil its requirements in the electronic industry (especially considering hybrid vehicles). Sea-mining offers a potential solution to the problem of mineral depletion. A comparison of potentially recoverable minerals from the brine of existing desalination plants and their current exploitation through conventional mining has been illustrated in Figure 2 [16]. It can be noted from the figure that the amount of Na and Mg in brine from current desalination capacities is more than that obtained through conventional mining. Similarly, strontium and lithium are both found considerable quantities in brine, indicating an attractive opportunity for the recovery of these minerals from the brine. The situation will further improve on the completion of contracted and planned desalination capacities across the globe.

In 2002, Drioli and co-workers [17] suggested for the first time membrane crystallizers for seawater desalination in an integrated approach with RO. In this study real seawater from the Tirrenian coast was first treated by nanofiltration (NF) and RO followed by membrane crystallization (MCr) treatment of the RO concentrate with the production of NaCl [17]. The prospect of introducing MCr to RO brine has the potential to increase the fresh water recovery factor from around 50% to above 90% in combination with salts recovery. In fact, one of the main advantages of MCr is that it does not suffer from osmotic phenomena like RO because the driving force is a temperature gradient instead of a pressure gradient. Therefore, MCr can treat highly concentrated and even saturated solutions without any particular reduction in driving force and, as a consequence, in trans-membrane flux. The hydrophobic character of the membrane provides a complete rejection of non-volatiles, thus producing water at very high qualities. Moreover, the low used temperatures (normally in the range of 40–80 °C) permit the use of waste grade heat or other renewable energy sources. In the subsequent years of this first study, several research activities have focused on integrated membrane systems in desalination. In 2004, a MCr unit was applied on a synthetic NF retentate. That resulted in the recovery of NaCl and magnesium sulfate in the form of Epsomite (MgSO$_4$·7H$_2$O) [18], a hydrated form of magnesium sulphate of great economic value. Another great advantage of membrane crystallizer is, in fact, the possibility to obtain different polymorphs and forms (hydrous or anhydrous) by simply tuning the operative conditions [19,20]. Effectively, in a MCr the membrane matrix acts as a selective gate for solvent evaporation, modulating the final degree and the rate for the generation of the supersaturation. Hence, the possibility to act on the trans-membrane flow rate, by changing the driving force of the process, allows for the potential to modulate the final properties of the crystals produced both in terms of structure (polymorphism) and morphology (habit, shape, size, and size distribution). Moreover, the possibility to separate the solvent evaporation (occurring inside the membrane module where the flowing solution is below the supersaturation condition) and the crystallization stage (performed in a separate tank on the retentate line operating in the metastable regime of supersaturation) allows for the proper control of the crystallization temperature, thus addressing the formation of a particular form (hydrous or anhydrous).

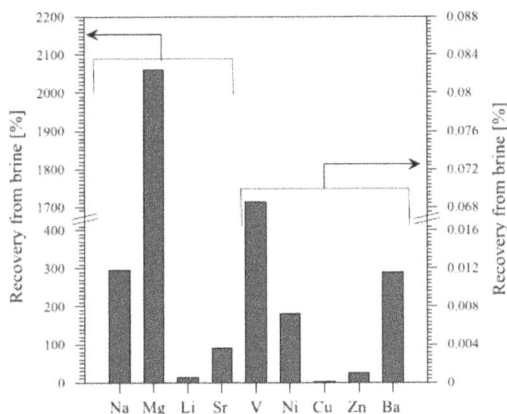

Figure 2. A comparative analysis of potentially recoverable minerals from brine in percentage with respect to their quantity obtained through traditional mining.

The European funded project: *Membrane Based Desalination: An Integrated Approach (MEDINA)* was launched in 2006. This project focused on integrated membrane systems for improving the efficiency of desalination. Integrated systems consisting of microfiltration (MF), NF, and RO with a membrane crystallizer coupled to NF and RO can achieve a water recovery factor as high as 92.4%, thus approaching zero liquid discharge and recovery of valuable raw materials [21]. Integration can be achieved in several configurations, as mentioned in [22–24]. Integrated membrane systems are also very interesting from an economical point of view. The specific water cost apparently increases when MD or MCr are introduced (Figure 3), mainly due to the requirement of steam when operating the thermal processes with respect to electrical energy demand in pressure driven membrane operations [24]. Nevertheless, the water recovery factor increases significantly with the introduction of MD and MCr from only 40% in RO (configuration 1) to above 90% for integrated operations (configuration 7). However, if the sale of CaCO$_3$, MgSO$_4$·7H$_2$O, and NaCl is considered the water production cost can be negative (solid lines shown in Figure 3) [24]. In this case, the desalinated fresh water can just be a by-product of mineral production. In addition to RO brine, other industrial saline water streams can also be considered for recovery of contained minerals.

Several developments are due for widespread applications of membrane distillation/crystallization process. The biggest issue is the unavailability of specific membranes with the required hydrophobicity, porosity, pore size distribution, and antifouling/wetting character. From a phenomenological point of view, a better understanding and control of temperature polarization is required. Research efforts are also required to improve the module design and to develop better logics for energy recovery. There is a need to develop large scale prototypes to assess the true techno-economic potential of the process and to comprehend the unforeseeable problems.

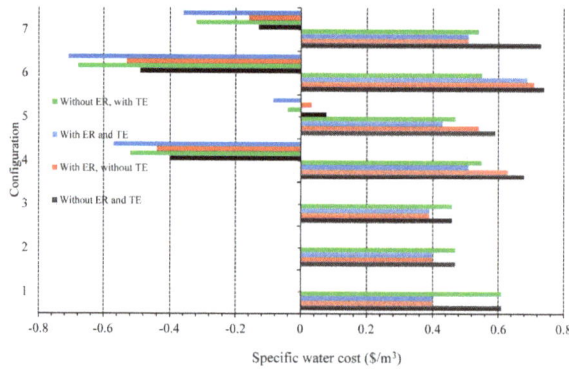

Figure 3. Specific water cost of different integrated configurations without considering sale of salt (patterned bars) and after considering sale of salt (solid lines). ER: Mechanical energy recovery; TE: Thermal energy recovery; MF, Microfiltration; NF, Nanofiltration; MD, membrane distillation; 1 = Configuration 1: RO alone; 2 = Configuration 2: NF-RO; 3 = Configuration 3: MF-NF-RO; 4 = Configuration 4: MF-NF(-MCr)-RO; 5 = Configuration 5: MF-NF-RO(-MCr); 6 = Configuration 6: MF-NF(-MCr)-RO(-MD); 7 = Configuration 7: MF-NF(-MCr)-RO(-MCr).

2.2. Energy Production in Desalination Sector

New developments and progresses in membrane engineering are creating exciting opportunities to generate renewable and sustainable energy in the desalination sector. This can be achieved by mixing two streams with different salinity gradients. The power generated is termed as blue energy or salinity gradient power. The total global potential of blue energy has been estimated to be about 1.4–2.6 TW out of which ~980 GW is extractable depending on the technology applied [25]. It has been demonstrated that salinity gradient energy is able to fulfil 20% of the current global energy demand. This can impart a significant contribution in lowering the dependence on carbon based energy production. PRO and RED are two of the most interesting membrane operations to harness this energy. Integration of these operations with conventional and emerging desalination operations produces synergetic effects. Besides reducing the net energy consumption of desalination processes, these operations also make desalination more clean and green by producing electricity with zero carbon emission and by diluting the concentrated brine which otherwise is a nuisance. By combining PRO or RED with membrane-based desalination systems like RO and MD, a synergetic advantage of both systems can be obtained in the logic of process intensification.

2.2.1. Reverse Electrodialysis (RED)

In hybrid RED-membrane desalination systems, highly concentrated reject brine (from both thermal and membrane desalination plants) is used for energy recovery. The generated electricity can be used to fulfill (partly or entirely) energy requirements of desalination system. Various integrations of RED with RO are possible [18]. RED can serve as a pretreatment, post treatment, or both steps for RO. Depending upon the configuration applied, RED can serve as energy reducer to the net energy generator in an RO desalination plant [18]. Similarly, integration of RED with MD can give synergetic effects. Application of MD at RO retentate can increase recovery factor while RED connected at retentate of MD can produce electricity while simultaneously decreasing its concentration.

Recently, Tufa et al. experimentally evaluated the integrated application of RED and MD for clean water and energy generation [22]. The conceptual illustration of the investigated hybrid system is presented in Figure 4. In particular, MD was operated with seawater RO retentate (1 M NaCl). Brine with concentrations of up to 5.4 M of NaCl (near supersaturation) was obtained from direct contact membrane distillation (DCMD) operated on RO brine (1 M NaCl) at a feed recirculation time of 20 h and a temperature gradient of 30 °C [15]. The DCMD brine was then used for energy recovery by RED system. The level of output power of RED depends on feed (MD brine) concentration and operating parameters like temperature and flow rate. For example, the maximum power density of a RED system was observed to increase from 0.9 W/m^2 to 2.4 W/m^2 when the MD brine concentration was increased from 4 M NaCl to 5.4 M NaCl, respectively [15]. The output power from the RED was also observed to increase with temperature (0.027 W/°C on average) and flow velocity, which can be optimized to set appropriate working conditions for improved system performance. In general, energy recovery from DCMD brine by RED enables the supply of extra energy required for desalination. This represents a promising strategy towards low energy desalination and Near-Zero Liquid Discharge.

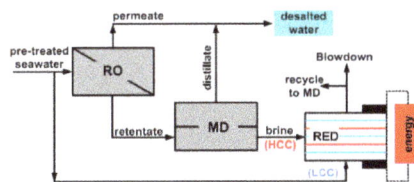

Figure 4. Integrated application of MD with RO and RED (reverse electrodialysis) for water and energy production [26]. HCC and LCC represent High Concentration Compartment and Low Concentration Compartment of the RED cell, respectively.

The feasibility of an integrated MD-RED depends on several factors affecting the performance of individual system units. Different strategies can be followed to improve the performance of the system in terms of water production capability, output power, and energy efficiency. The energy consumption in RO or MD can be reduced by controlling the operating conditions, using appropriate pre-treatment techniques, and designing optimal membranes and modules [12,21]. The potential and hence the output power from a RED unit can be improved by the development and use of highly permeable and low resistance IEMs (ion exchange membranes) [11]. Besides the benefit obtained at higher temperatures for both systems (MD and RED), the coupled system enables the possibility to use and convert waste heat to electricity.

2.2.2. Pressure Retarded Osmosis (PRO)

PRO is one of the most interesting membrane processes to harness clean energy from a salinity gradient. In PRO, a semipermeable membrane is applied to separate high and low salinity solutions. The osmotic pressure extracts the fresh water from dilute to the concentrated solution. The pressure generated on the high salinity solution side can be used to run a hydro turbine [27,28]. For this process, several combinations of feed and draw solutions have been tested including river and seawater [29], seawater brine and wastewater retentate, freshwater and synthetic NaCl solution [30], and seawater and municipal wastewater [31].

Synergetic effects can be achieved by introducing PRO into desalination systems in terms of reduction of waste footprint of feed solution, and the dilution of the draw solution, which can reduce the energy demand of desalination system and can minimize the environmental consequences of brine disposal. Achili et al. [31] have reported experimental data on a RO unit getting a benefit from pressure generated by a PRO unit. The authors have also reported the power density data of PRO-RO system ranging between 1.1 W/m^2 to 2.3 W/m^2. The authors have concluded that the proposed system can bring the desalination energy demand down to 1 kWh/m^3. In order to increase the power density of the PRO process, Han et al. [30] have proposed the integration of PRO with a closed loop

MD process. MD maintains a high concentration of draw solution that extracts freshwater from the feed solution. The authors have claimed that the proposed system can achieve a high recovery factor, huge production of power, and minimum membrane fouling and environmental impacts.

The concept of integrating PRO with other membrane operations is also gaining attention at industrial scales. Currently, Applied Biomimetics and partners have setup two pilot plants aiming to generate electricity from geothermal wastewater by using PRO technology (energyforskining.dk/node/8345). The proposed plan is expected to produce emission free electricity while at the same time bringing the salinity level of the geothermal brine below the permissible limits. The dilution of geothermal brine via PRO will also reduce the corrosion and scaling potential of a geothermal stream. A similar concept has been used in a Mega-ton project where a pilot plant has been constructed in Fukuoka (Japan). The plant uses 460 m^3/day of RO brine which is mixed with 420 m^3/day of wastewater. The plant has been able to achieve power density as high as 13 W/m^2 at 30 bar hydraulic pressure by using commercial hollow fibers from TOYOBO (Osaka, Japan).

Another pilot-scale PRO-hybrid research project is being conducted under the name "Global MVP (Membrane distillation, Valuable resource recovery, Pressure retarded osmosis) Project" in Korea (Figure 5). The objective of this project was to evaluate the feasibility of the RO-MD-PRO hybrid process in terms of reducing the discharged water concentration and the energy consumption. In the hybrid process, the concentrated RO brine enters the MD feed side, and the further concentrated MD brine is then utilized as a PRO draw solution while the waste water effluent is used as the feed solution. Consequently, an improvement in the total plant efficiency compared to a stand-alone RO plant is expected due to the additional water production by MD and the reduction of net energy consumption resulting from the PRO energy generation. Specifically, the following pilot plant will be built: a RO system capable of 1000 m^3/day water production, a MD system with a water production capacity of 400 m^3/day, and a PRO system having a 5 W/m^2 power density [32]. Recently Statkraft has terminated its activities on power generation through PRO. The company has been operating a prototype with 10 kW capacity by applying seawater and river water as draw and feed solutions, respectively, which were separated through a membrane with power density of 1 W/m^2, which was far less than the economical break-even point (5 W/m^2). Although membranes with power densities as high as 10 W/m^2 have been reported in current literature, the price and commercial availability of these membranes still remain the unanswered questions. Besides high power density, the appropriate membranes should exhibit high selectivity and minimum reverse solute diffusion. Internal concentration polarization and fouling are the other main issues hindering the performance of PRO processes. Due to exposure of the support layer to the feed solution, fouling and internal concentration polarization issues are more severe in PRO than pressure driven processes. Besides the proper treatment of feed solution, the design and modification of the support layer must be emphasized to alleviate this phenomenon.

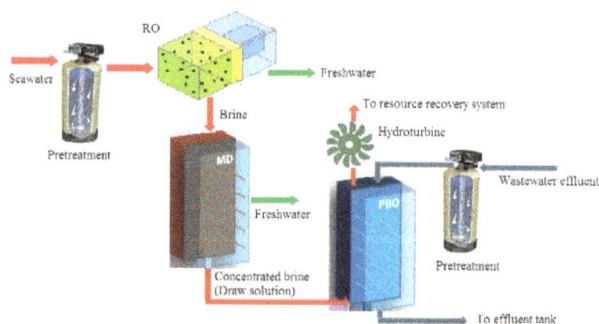

Figure 5. Schematic diagram of hybrid RO-MD-PRO in Korea implemented under Global MVP project.

3. Evaluation of New Metrics

A comparison of different overall desalination processes in terms of various metrics for process intensification has been provided in Figure 6. MSF (multi-stage flash), being the most widely used thermal desalination process, has been considered as the base line and therefore the value of all the metrics for this is one. The figure indicates that various metrics show huge variations for different processes. MI (mass intensity) for RO and MD reduce greatly due to improved recovery factor (considered 50% and 86%, respectively, in current study). RO shows the maximum value of productivity/weight ratio (PW) due to high membrane permeability and elimination of heavy metallic parts, which are essential components of MSF plant. This aspect is particularly important for off-shore or remote installations. It is also evident from the figure that the PW for RO is higher than MD, which can be attributed toward the high flux of RO considered in the current study (Table 1). Larger membrane area requirement in case of MD implies a larger number of modules that will increase the weight, and thus PW will go down. Similarly, PS for MSF is the least, thereby indicating the large foot print of the plant. Overall, the comparison indicates that new membrane operations (MD in the current example) can be optimum candidates to overcome the drawbacks (limited MI) of conventional RO processes.

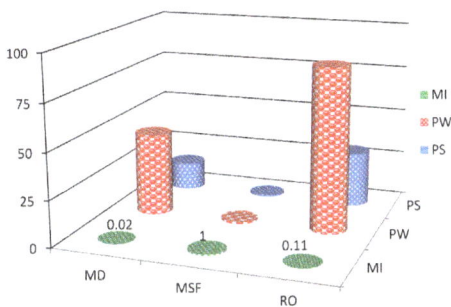

Figure 6. PI (process intensification) metrics for MD, multi stage flash (MSF), and RO. MI, mass intensity; PW, productivity/weight ratio; PS, productivity/size ratio.

Table 1. Parameters and assumption used in calculation of various PI (process intensification) metrics shown in Figure 6.

Feed solution	Seawater
Plant capacity (m³/h)	1250 [33]
MD flux assumed (kg/m²·h)	4
Volume of MSF (multi stage flash) unit (m³)	18 × 4 × 3 [33]
Material of construction of MSF unit	Stainless steel
Density of steel (kg/m³)	8000
Average permeate flow for RO (reverse osmosis) (m³/h)	24.6
Weight of one element for RO/MD (kg)	16
volume of one RO/MD element (m³)	1.016 × 0.0286 × 0.201

Weight and volume of MD module has been considered equal to that of the RO reported in [34].

Besides the traditional low concentrated solutions, emerging membrane operations have the capability to treat highly concentrated solutions which are beyond the application limit of traditional processes, thus offering the possibility to achieve the values of MI and WI (Waste intensity) which otherwise are not feasible. This aspect has been explained by considering the example of the treatment of produced water by using membrane crystallization. The experimental details of the study have been explained elsewhere [35]. Changes of mass and waste intensities and overall recovery with respect to size of membrane (0.2 m^2—active surface area) and weight (0.467 kg—module) were identified (Figure 7). In the beginning of the experiment, water was only considered the product and in the end of the experiment both water and salt were considered. From the time of saturation, NaCl was being produced at a rate of 0.063 $\pm$ 0.012 kg/h. MI and WI decreased significantly with increases in water recovery factor from above 35 to below 3 and 2 for MI and WI, respectively. In the carried out experiments a recovery factor of only 37% was obtained and continued treatment could further improve MI and WI.

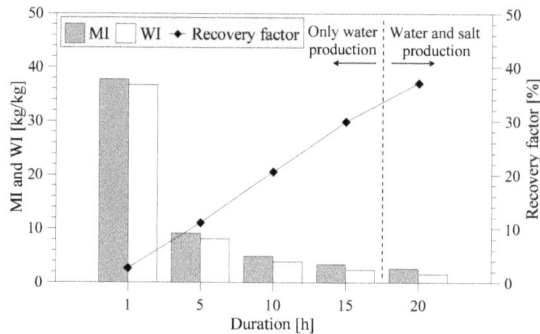

Figure 7. Mass and waste intensities and overall water recovery factor with duration of treatment.

4. Renewable Energy in Desalination

Despite rapid increases in membrane based desalination facilities, the use of fossil fuel based energy creates serious concerns regarding the sustainability and cost of obtaining water from desalination. In order to make desalination perfectly aligned with objectives of PIS, the carbon footprint of desalination plants must be reduced. Furthermore, the energy consumption of the process must be lowered to make it affordable for less fortunate communities as well. So far, the largest expansion of desalination facilities has been recorded in energy rich regions and/or regions of the developed world including Spain, Australia, and North America. This is due to the high energy demand of traditional desalination processes. Reverse osmosis is the most economical desalination technology available at a commercial scale at the moment, but still the energy consumption of this technology is more than double the minimum theoretical energy requirements of the process [36]. This scenario has attracted a large interest in the development of desalination based on renewable energy resources.

The cost of renewable desalination is in the same range as that for the traditional resources [37]. The types of renewable energy sources suitable for desalination mainly include solar energy, geothermal energy, and wind energy. The selection of renewable energy sources and applied desalination techniques (RO, thermal, EDs (electrodialysis), etc.) depends upon several factors, mainly including the type of renewable energy sources available in the area, the nature of local water (seawater or brackish water), production capacity, remoteness of the area, etc. [38]. Currently, most of the renewable based desalination facilities are based upon RO and are driven by solar and wind energy [39]. These plants have capacities ranging from a few cubic meters to several hundred cubic meters per day. The specific energy consumption of these units for brackish water ranges from 0.9 to 29.1 kWh/m^3 and 2.4 to 17.9 kWh/m^3 for brackish and seawater, respectively [40]. Solar and wind energy sources, however, do not allow continuous operation without the use of storage batteries. To overcome this drawback, renewable energy desalination has been integrated with conventional energy sources including grid electricity, diesel generators, etc. The second largest beneficial form of renewable energy based desalination is electrodialysis, which has mainly been operated with wind energy. MD represents another emerging player in the field. MD plants driven with solar energy and geothermal energy have been installed in different regions across the globe.

Membrane based renewable desalination is expected to emerge as an interesting sector in the future to reduce carbon emission and desalination costs, particularly in remote regions with abundant sources of saline water and renewable energy but poor infrastructure of electricity. The key factor for further growth of renewable energy desalination is the development and progress in more efficient, cheap renewable energy systems, and improved control and logic systems to regulate the energy output. Moreover, the techno-economic feasibility of membrane based desalination technologies driven with renewable energy sources at large scale needs to be investigated.

5. Conclusions and Perspectives

The concept of the process intensification strategy has been significantly implemented in the desalination sector where large, heavy, and energy intensive technologies have been replaced with membrane based technologies with low energy consumption, small footprints, and low environmental consequences. On the basis of process intensification metrics, the membrane operations clearly surpass their conventional thermal counterparts. Relatively less explored membrane based operations including MD/MCr, PRO, RED, etc. are emerging with demonstrated potential to increase water recovery factor, to achieve minerals recovery and renewable energy production, thus breaking the traditional bounds and applications of membrane technology in desalination. Interesting developments of these operations are appearing at commercial scales and mega desalination projects including MEDINA, Seahero, Mega-ton, and the Global MVP project have witnessed the interest of practical applications of these processes. The reshaping of the desalination industry can be foreseen due to the implementation of these processes, and it can be realistically expected that the desalination industry will turn into the energy and raw material sector in the future, with freshwater as a useful byproduct. The promises of new processes, however, are strongly related with successfully overcoming several challenges including the development of better membranes for PRO, RED, and MD/MCr, better control of internal concentration polarization and biofouling, better spacer design, improvement in efficiency of energy recovery devices, improved module design, etc. In remote areas with poor infrastructure, the implementation of renewable energy sources in membrane based desalination is expected to gain more attention. MD is expected to emerge as a strong contester of RO in this field, particularly in the regions with abundant sources of solar and geothermal energy.

Author Contributions: Aamer Ali performed the simulation analysis and wrote the manuscript. Francesca Macedonio supervised the work and revised the manuscript. Enrico Drioli initiated the work.

Conflicts of Interest: The authors declare no conflict of interest.

Abbreviations

The following symbols are used in this manuscript:

Configuration 1	RO alone
Configuration 2	NF-RO
Configuration 3	MF-NF-RO
Configuration 4	MF-NF(-MCr)-RO
Configuration 5	MF-NF-RO(-MCr)
Configuration 6	MF-NF(-MCr)-RO(-MD)
Configuration 7	MF-NF(-MCr)-RO(-MCr)
DCMD	Direct contact membrane distillation
ED	Electrodialysis
ER	Mechanical energy recovery
FO	Forward osmosis
IEMs	Ion exchange membranes
MCr	Membrane crystallization

MD	Membrane distillation
MD-RED	Membrane distillation—reverse electrodialysis
MI	Mass intensity
MEDINA	Membrane Based Desalination: An Integrated Approach
MF	Microfiltration
MSF	Multi-stage flash
NF	Nanofiltration
PIS	Process intensification strategy
PS	Productivity/size ratio
PW	Productivity/weight ratio
PRO	Pressure retarded osmosis
RO	Reverse osmosis
RED	Reverse electrodialysis
TE	Thermal energy recovery
WI	Waste intensity

References

1. Randers, J. 2052: A Global Forecast for the Next Forty Years. Available online: http://www.2052.info/ (accessed on 8 October 2016).
2. UN. Coping with Water Scarcity—Challenge of the Twenty-First Century. Available online: http://www.unwater.org/wwd07/downloads/documents/escarcity.pdf (accessed on 15 October 2016).
3. Stankiewicz, A.I.; Moulijn, J.A. Process Intensification: Transforming Chemical Engineering. *Chem. Eng. Prog.* **2000**, *96*, 22–34. [CrossRef]
4. Stankiewicz, A. Reactive separations for process intensification: An industrial perspective. *Chem. Eng. Process.* **2003**, *42*, 137–144. [CrossRef]
5. Drioli, E.; Brunetti, A.; di Profio, G.; Barbieri, G. Process intensi Fi cation strategies and membrane engineering. *Green Chem.* **2012**, *14*, 1561–1572. [CrossRef]
6. Drioli, E.; Giorno, L. (Eds.) *Comprehensive Membrane Science and Engineering*; Elsevier B.V.: Amsterdam, The Netherlands, 2010.
7. Charpentier, J.-C. Among the trends for a modern chemical engineering, the third paradigm: The time and length multiscale approach as an efficient tool for process intensification and product design and engineering. *Chem. Eng. Res. Des.* **2010**, *88*, 248–254. [CrossRef]
8. Drioli, E.; Stankiewicz, A.I.; Macedonio, F. Membrane engineering in process intensification—An overview. *J. Membr. Sci.* **2011**, *380*, 1–8. [CrossRef]
9. Macedonio, F.; Drioli, E.; Gusev, A.A.; Bardow, A.; Semiat, R.; Kurihara, M. Efficient technologies for worldwide clean water supply. *Chem. Eng. Process. Process Intensif.* **2012**, *51*, 2–17. [CrossRef]
10. Global Water Intelligence. *IDA Desalination Yearbook 2016–2017*; Media Analytics, Ltd.: Oxford, UK, 2016.
11. Zhu, A.; Christofides, P.D.; Cohen, Y. Effect of Thermodynamic Restriction on Energy Cost Optimization of RO Membrane Water Desalination. *Ind. Eng. Chem. Res.* **2009**, *48*, 6010–6021. [CrossRef]
12. Zhu, A.; Christofides, P.D.; Cohen, Y. On RO membrane and energy costs and associated incentives for future enhancements of membrane permeability. *J. Membr. Sci.* **2009**, *344*, 1–5. [CrossRef]
13. Zhu, A.; Christofides, P.D.; Cohen, Y. Energy Consumption Optimization of Reverse Osmosis Membrane Water Desalination Subject to Feed Salinity Fluctuation. *Ind. Eng. Chem. Res.* **2009**, *48*, 9581–9589. [CrossRef]
14. Qi, W.; Liu, J.; Christofides, P.D. Supervisory Predictive Control for Long-Term Scheduling of an Integrated Wind/Solar Energy Generation and Water Desalination System. *IEEE Trans. Control Syst. Technol.* **2012**, *20*, 504–512. [CrossRef]
15. Criscuoli, A.; Drioli, E. New Metrics for Evaluating the Performance of Membrane Operations in the Logic of Process Intensification. *Ind. Eng. Chem. Res.* **2007**, *46*, 2268–2271. [CrossRef]
16. Quist-Jensen, C.A.; Macedonio, F.; Drioli, E. Membrane crystallization for salts recovery from brine—An experimental and theoretical analysis. *Desalin. Water Treat.* **2015**, *3994*, 1–11. [CrossRef]
17. Drioli, E.; Criscuoli, A.; Curcio, E. Integrated membrane operations for seawater desalination. *Desalination* **2002**, *147*, 77–81. [CrossRef]

18. Drioli, E.; Curcio, E.; Criscuoli, A.; di Profio, G. Integrated system for recovery of CaCO₃, NaCl and MgSO₄·7H₂O from nanofiltration retentate. *J. Membr. Sci.* **2004**, *239*, 27–38. [CrossRef]

19. Di Profio, G.; Tucci, S.; Curcio, E.; Drioli, E. Selective Glycine Polymorph Crystallization by Using Microporous Membranes. *Cryst. Growth Des.* **2007**, *7*, 526–530. [CrossRef]

20. Quist-jensen, C.A.; Ali, A.; Mondal, S.; Macedonio, F.; Drioli, E. A study of membrane distillation and crystallization for lithium recovery from high-concentrated aqueous solutions. *J. Membr. Sci.* **2016**, *505*, 167–173. [CrossRef]

21. Drioli, E.; Criscuoli, A.; Macedonio, F. (Eds.) *Membrane Based Desalination: An Integrated Approach (Medina) (European Water Research)*; IWA Publishing: London, UK, 2011.

22. Macedonio, F.; di Profio, G.; Curcio, E.; Drioli, E. Integrated membrane systems for seawater desalination. *Desalination* **2006**, *200*, 612–614. [CrossRef]

23. Drioli, E.; Curcio, E.; di Profio, G.; Macedonio, F.; Criscuoli, A. Integrating Membrane Contactors Technology and Pressure-Driven Membrane Operations for Seawater Desalination. *Chem. Eng. Res. Des.* **2006**, *84*, 209–220. [CrossRef]

24. Macedonio, F.; Curcio, E.; Drioli, E. Integrated membrane systems for seawater desalination: Energetic and exergetic analysis, economic evaluation, experimental study. *Desalination* **2007**, *203*, 260–276. [CrossRef]

25. Post, J.W.; Goeting, C.H.; Valk, J.; Goinga, S.; Veerman, J.; Hamelers, H.V.M.; Hack, P.J.F.M. Towards implementation of reverse electrodialysis for power generation from salinity gradients. *Desalin. Water Treat.* **2010**, *16*, 182–193. [CrossRef]

26. Ashu, R.; Curcio, E.; Brauns, E.; van Baak, W.; Fontananova, E.; Di, G. Membrane Distillation and Reverse Electrodialysis for Near-Zero Liquid Discharge and low energy seawater desalination. *J. Membr. Sci.* **2015**, *496*, 325–333.

27. Wang, X.; Huang, Z.; Li, L.; Huang, S.; Hao, E.; Scott, K. Energy Generation from Osmotic Pressure Difference Between the Low and High Salinity Water by Pressure Retarded Osmosis. *J. Technol. Innov. Renew. Energy* **2012**, *1*, 122–130. [CrossRef]

28. Klaysom, C.; Cath, T.Y.; Depuydt, T.; Vankelecom, I.F.J. Forward and pressure retarded osmosis: Potential solutions for global challenges in energy and water supply. *Chem. Soc. Rev.* **2013**, *42*, 6959–6989. [CrossRef] [PubMed]

29. O'Toole, G.; Jones, L.; Coutinho, C.; Hayes, C.; Napoles, M.; Achilli, A. River-to-sea pressure retarded osmosis: Resource utilization in a full-scale facility. *Desalination* **2016**, *389*, 39–51. [CrossRef]

30. Han, G.; Zuo, J.; Wan, C.; Chung, T. Hybrid pressure retarded osmosis—Membrane distillation (PRO—MD) process for osmotic power and clean water generation. *Environ. Sci. Water Res. Technol.* **2015**, *1*, 507–515. [CrossRef]

31. Achilli, A.; Prante, J.L.; Hancock, N.T.; Maxwell, E.B.; Childress, A.E. Experimental Results from RO-PRO: A Next Generation System for Low-Energy Desalination. *Environ. Sci. Technol.* **2014**, *48*, 6437–6443. [CrossRef] [PubMed]

32. Global MVP 2013–2018. Available online: http://globalmvp.org/english/ (accessed on 16 October 2016).

33. El-Dessouky, H.T.; Ettouney, H.M. *Fundamentals of Salt Water Desalination*; Elsevier B.V.: Amsterdam, The Netherlands, 2002; pp. 271–407.

34. GE Water & Process Technologies, AD HR Series, Seawater RO High Rejection Elements. Available online: http://www.lenntech.com/Data-sheet/GE-Osmonics-AD-HR-series-Sea-Water-RO-High-Rejection-Desalination.pdf (accessed on 27 Otober, 2016).

35. Ali, A.; Quist-jensen, C.A.; Macedonio, F.; Drioli, E. Application of Membrane Crystallization for Minerals' Recovery from Produced Water. *Membranes* **2015**, *5*, 772–792. [CrossRef] [PubMed]

36. Khayet, M. Solar desalination by membrane distillation: Dispersion in energy consumption analysis and water production costs (A review). *Desalination* **2013**, *308*, 89–101. [CrossRef]

37. Gnaneswar, V.; Nirmalakhandan, N.; Deng, S. Renewable and sustainable approaches for desalination. *Renew. Sustain. Energy Rev.* **2010**, *14*, 2641–2654.

38. Eltawil, M.A.; Zhengming, Z.; Yuan, L. Renewable energy powered desalination systems: Technologies and economics-state of the art. In Proceedings of the 12th International Water Technology Conference, Alexandria, Egypt, 27–30 March 2008; pp. 1–38.

39. Tzen, E.; Morris, R. Renewable energy sources for desalination. *Sol. Energy* **2003**, *75*, 375–379. [CrossRef]
40. Ghermandi, A.; Messalem, R. Solar-driven desalination with reverse osmosis: The state of the art. *Desalin. Water Treat.* **2009**, *7*, 285–296. [CrossRef]

applied
sciences

MDPI

Article

Improving Nanofiber Membrane Characteristics and Membrane Distillation Performance of Heat-Pressed Membranes via Annealing Post-Treatment

Minwei Yao, Yun Chul Woo, Leonard D. Tijing *, Cecilia Cesarini and Ho Kyong Shon *

Centre for Technology in Water and Wastewater, School of Civil and Environmental Engineering, University of Technology Sydney (UTS), 15 Broadway, Ultimo NSW 2007, Australia; minwei.yao@student.uts.edu.au (M.Y.); Yunchul.Woo@student.uts.edu.au (Y.C.W.); Cecilia.Cesarini@ uts.edu.au (C.C.)
* Correspondence: Leonard.Tijing@uts.edu.au (L.D.T.); Hokyong.Shon-1@uts.edu.au (H.K.S.); Tel.: +61-2-9514-2652 (L.D.T.); +61-2-9514-2629 (H.K.S.)

Academic Editor: Enrico Drioli
Received: 16 December 2016; Accepted: 5 January 2017; Published: 12 January 2017

Abstract: Electrospun membranes are gaining interest for use in membrane distillation (MD) due to their high porosity and interconnected pore structure; however, they are still susceptible to wetting during MD operation because of their relatively low liquid entry pressure (LEP). In this study, post-treatment had been applied to improve the LEP, as well as its permeation and salt rejection efficiency. The post-treatment included two continuous procedures: heat-pressing and annealing. In this study, annealing was applied on the membranes that had been heat-pressed. It was found that annealing improved the MD performance as the average flux reached 35 $L/m^2 \cdot h$ or LMH (>10% improvement of the ones without annealing) while still maintaining 99.99% salt rejection. Further tests on LEP, contact angle, and pore size distribution explain the improvement due to annealing well. Fourier transform infrared spectroscopy and X-ray diffraction analyses of the membranes showed that there was an increase in the crystallinity of the polyvinylidene fluoride-*co*-hexafluoropropylene (PVDF-HFP) membrane; also, peaks indicating the α phase of polyvinylidene fluoride (PVDF) became noticeable after annealing, indicating some β and amorphous states of polymer were converted into the α phase. The changes were favorable for membrane distillation as the non-polar α phase of PVDF reduces the dipolar attraction force between the membrane and water molecules, and the increase in crystallinity would result in higher thermal stability. The present results indicate the positive effect of the heat-press followed by an annealing post-treatment on the membrane characteristics and MD performance.

Keywords: membrane distillation; post-treatment; annealing; PVDF-HFP; crystallinity

1. Introduction

The shortage of water is one of the biggest concerns for future society as the human population is increasing steadily. Desalination, a good option for coastal areas short of fresh water, has been becoming the major approach for potable water as the supply of seawater can be considered unlimited. Presently, reverse osmosis (RO) has reached the state of the art and has become the dominant technology because of its higher energy efficiency and stability compared to conventional thermal-based processes [1]. However, there are two main issues in RO: high energy consumption and brine treatment, which are becoming great challenges for future human society [2]. Hence, continuous efforts for finding new technologies that can provide lower energy consumption while still obtaining high process and production efficiencies are being sought out [3]. Among them, membrane distillation (MD) is one of the most promising emerging technologies [4].

One of the major advantages for MD is the potential usage of low-grade waste heat as its feed temperature requirement is much lower than that of the conventional distillation process. If sufficient waste heat is available, much less energy will be required for operation. Fundamentally, the membranes in MD serve as contactors, and are not involved in the separation process themselves. The mass transport starts from the evaporation of water at the boundary between the vapor and liquid phase at the membrane pores. Then the vapor is driven by the partial pressure, caused by a partial vapor pressure difference which is triggered by the temperature difference between the hot feed and cold permeate [5,6].

Although MD has these unique advantages, major challenges must still be addressed for its wide acceptance in the industry, which include a lack of specifically designed MD membranes and modules, difficulty in up-scaling the laboratory setup, and a shortage of techno-economic data [6]. Membrane fabrication has gained a great deal of interest, as researchers are using membranes designed for other processes. However, the characteristics of a decent MD membrane are different, as much higher hydrophobicity and liquid entry pressure (LEP) with high porosity are needed for a high flux performance with minimal wetting issues [7]. Polymers with relatively lower surface energies, such as polystyrene, polyvinylidene and polytetrafluoroethylene, have been used to fabricate membranes in the laboratory [8–11]. Phase inversion is one of the most used techniques due to its simplicity [12,13]. However, its products usually have low flux in MD due to its relatively low porosity and pore size [14,15].

Currently, electrospinning is gaining popularity in membrane fabrication [8–10,13,16]. Electrospun membranes have many advantages including a high contact angle, very high porosity, and simplicity for modification, making them very suitable for MD [17,18]. However, they have a large maximum pore size and, hence, a low LEP, making them susceptible to wetting [6,18]. Much effort has been invested into finding solutions to improve their LEP. For example, Prince fabricated a modified electrospun membrane by adding lab-made macromolecules into the solution, and later made a composite membrane consisting of both electrospun and phase inversion membranes, improving its LEP enormously [13,19]. Liao improved the LEP by adding surface-modified silica nanoparticles into the solution [16]. Lee added fluorosilane-coated TiO_2 into the solution and obtained an electrospun membrane with an increased LEP [20]. Other hydrophobic nanoparticles used as additives, including graphene and carbon nanotubes, were also extensively studied [12,21]. The other method is to incorporate a secondary polymer with very low surface tension into the polymer solution. Polydimethylsiloxane (PDMS) was found to be able to improve the hydrophobicity and LEP when mixed with the carrier polymer in the solution [22,23].

Post-modification techniques such as heat-pressing have been studied to improve the characteristics and membrane distillation performance of the electrospun membranes [24,25]. In our previous study, the effects of heat-press conditions were fully studied [24]. It was found that the temperature and duration played more important roles than pressure during heat-pressing. The properties of the electrospun membrane were impressively improved by heat-pressing. Although the contact angle and porosity decreased, the LEP and, hence, permeation performance in desalination improved greatly. In this study, to further improve the properties of the electrospun membranes, annealing, a commonly practiced thermal treatment, was applied to the heat-pressed membrane. A favorable change of the properties was achieved, and the MD performance was further improved.

2. Materials and Methods

2.1. Materials

Polyvinylidene fluoride-*co*-hexafluoropropylene (referred herein as PVDF-HFP, MW = 455,000) was purchased from Sigma-Aldrich, Sheboygan, WI, USA. For membrane fabrication, acetone (ChemSupply, Adelaide, Australia) and N,N-Dimethylacetamide (DMAc, Sigma-Aldrich, Sheboygan, WI, USA) were utilized as solvents. All the chemicals were used as received without further purification. A polypropylene (PP) filter layer purchased from Ahlstrom (Helsinki, Finland) was applied as support layer in all the MD tests except when commercial membrane was in use. Commercial microfiltration membrane (pore size = 0.22 μm, porosity = 70%, GVHP) bought from Millipore (Jaffrey, NH, USA) was tested for comparison.

2.2. Membrane Fabrication by Electrospinning

PVDF-HFP (20 wt. %) was dissolved in a composite solvent comprising acetone and DMAc (1:4 acetone/DMAc ratio). The polymer powder was added into the mixed solvent and stirred by magnetic stirrer for 24 h at room temperature for complete dissolution. During electrospinning process, a 6 mL volume of the polymer solution was electrospun at a rate of 1 mL/h by applying 21 kV voltage between the tip of the spinneret and the rotating collector (metal drum) with a tip-to-collector distance of 20 cm. The relative humidity of the process chamber during electrospinning was in the range of 46%–54% at room temperature.

2.3. Post-Treatment of Electrospun Membranes

After the completion of the electrospinning, the just-fabricated membranes were removed from the collector and dried at 50 °C for 2 h inside an air flow oven (OTWMHD24, Labec, Sydney, Australia). The membranes were then heat-pressed by being set between flat metal plates with dead weight (6.5 kPa) placed on the top plate in a pre-heated oven at temperature of 150 °C, while fully covered by foils. A 24 h heat-pressing was implemented for thorough microstructure evolution.

After heat-pressing, membrane annealing was realized by removing the dead weight on the membranes and leaving the pressed membranes in the oven at 120 °C (where temperature had been gradually decreased by 10 °C per hour from 150 °C) for another one to three days. After these day(s), the membranes were slowly cooled by reducing temperature in the oven by 10 °C per hour, until room temperature was reached. The samples were named in Table 1 as shown below.

Table 1. Sample names and thicknesses of the membranes prepared in the present study.

Sample Name	Description	Membrane Thickness (μm)
Neat	As-spun electrospun PVDF-HFP (polyvinylidene fluoride-*co*-hexafluoropropylene) membranes	51
HP	Neat membrane heat-pressed at 150 °C under 6.5 kPa for 24 h	39
A1	HP membranes annealed for 1 day at 120 °C	37
A2	HP membranes annealed for 2 days at 120 °C	34
A3	HP membranes annealed for 3 days at 120 °C	34

2.4. Characterization

Contact angle, which was generally used to indicate hydrophobicity of membrane surface, was measured by Theta Lite 100 (Attension) (Biolin Scientific, Paramus, NJ, USA) with sessile drop method [24,26]. Then 5~8 μL of water droplet was placed onto the membrane surface for analysis. A mounted motion camera was applied to capture the images at a rate of 12 frames per second. Contact angle could be obtained by analyzing the recorded video with aid of the Attension software. An average of three values was used as contact angle data for each sample.

As-spun, heat-pressed and annealed membrane samples was measured with a lab-made setup for their liquid entry pressure [9,24]. A gas supply was connected to a hollow stainless plate by a tube, with a digital gauge standing as the intermediate between them. On the top of the stainless plate, a stainless cylinder container was fully filled with distilled water while a plastic plug stuck in its bottom. The samples were firmly fixed on the top of cylinder by a stainless cap with a lock catch. The nitrogen gas was steadily released to increase the pressure by 5 kPa per 30 s, until the first bubble sign appeared, which was recorded as LEP. Each sample was tested in triplicate and average data was recorded.

Fourier-transform infrared (FT-IR) spectroscopy (Varian 2000, Agilent, Santa Clara, CA, USA) was used to investigate the PVDF-HFP phases of the virgin membrane and thermally-treated membranes. Each spectrum was acquired with signal averaging 32 scans at a resolution of 8 cm^{-1}, in transfer mode by pressing the sample with KBr to a pellet.

X-ray diffraction (XRD) (Siemens D5000, Siemens, Karlsruhe, Germany) was carried out over Bragg angles ranging from $10°$ to $30°$ (Cu Kα, λ = 1.54059 Å).

The pore size and pore size distribution of the fabricated neat and post-treated electrospun membranes were measured by capillary flow porometry (CFP-1200-AEXL, Porous materials Inc., Ithaca, NY, USA). All samples were firstly wetted by Galwick (a wetting liquid with a low surface tension of 15.9 dynes/cm) and tested under the pre-set conditions. Then the dried samples were applied with N_2 gas to determine the gas permeability under same conditions. The final average pore size and its distribution were automatically calculated with both data sets of wet and dry tests by the specific software.

Membrane porosity was calculated by using gravimetric method, where the volume of the membrane pores was divided by the total volume of the whole membrane. Ethanol (Univar 1170 from Ajax Finechem Pty Ltd., Sydney, Australia) was used to completely wet the membranes. The weight (w_1, g) of wetted membrane was measured after the residual ethanol on the surface was removed. Then the membrane samples were dried after being left still in the open air for 15 min and then weighed (w_2, g). The porosity of the electrospun membranes could be calculated with the following equation:

$$\varepsilon_m = \frac{(w_1 - w_2)/\rho_e}{(w_1 - w_2)/\rho_e + w_2/\rho_p}$$

where ρ_e is the density of the ethanol (g/m^3) and ρ_p is the density of the PVDF-HFP (g/m^3) [16,24].

2.5. Direct Contact Membrane Distillation (DCMD) Test

MD has several common configurations: direct contact MD (DCMD), air gap MD (AGMD), vacuum MD (VMD), and sweeping gas MD (SGMD) [2]. Recently, permeate gap MD (PGMD) is gaining lots of interests due to its simple configuration [6], this study was focused on DCMD configuration, which is illustrated in Figure 1. Supported by a PP filter layer, the membrane samples were fixed in the DCMD cell module with a length and width of 77 mm and 26 mm, respectively, making up an effective membrane area of 20 cm^2, for both feed and permeate channels. The module was placed horizontally and ran in counter-current mode with feed flow on top side [9]. Sodium chloride (NaCl) (3.5 wt. % concentration) was used as feed, and deionized (DI) water was used as permeate, with temperature maintained at 60 °C and 20 °C, respectively. The mass flow rates of 400 mL/min were maintained by gear pumps for both feed and permeate flows. A desktop computer was used to collect the data of mass of permeate tank and temperatures in both feed and permeate tanks automatically. Permeation performances (flux and salt rejection) of post-treated electrospun membranes were compared with a commercial membrane (GVHP, 0.22 pore size, and 110 µm thickness).

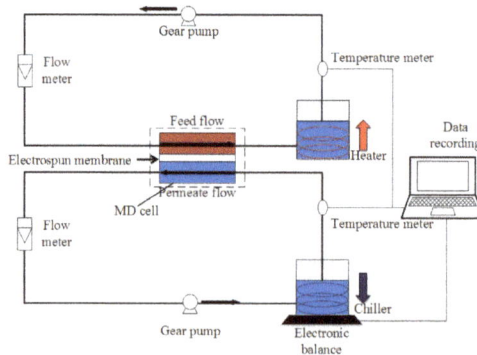

Figure 1. Schematic figure of DCMD (direct contact membrane distillation) system used in this study.

3. Results and Discussion

3.1. DCMD Performance

All the heat-pressed and annealed membranes had a salt rejection of 99.99%, while the neat membrane suffered rapid wetting immediately 30 s after the operation started. The wetting of electrospun membranes could be judged based on the steady increase in the flux and the conductivity of the permeate (exceeding 10 mS/cm). Figure 2 shows that optimal heat-pressing improves the flux and wetting resistance. An average of 28.7 LMH was obtained during 10 h of operation, which is much higher than in previous studies under similar conditions. For comparison, Liao et al. achieved a flux of 20.6 LMH with heat-pressed electrospun nanofiber membranes in DCMD [27]. Another study obtained a flux of 22 LMH with heat-pressed, two-layer membranes where the PVDF-HFP concentration was 10% [28]. However, the heat-pressed electrospun nanofiber membrane in the present study still has a noticeable decreasing trend during the 10 h of operation. On the other hand, annealed membranes showed a noticeably higher flux than the heat-pressed membrane, although they shared a close membrane thickness (Figure 2). Moreover, the annealed membrane showed a more stable trend of flux in 10 h of operation than HP, indicating a better wetting resistance. The membrane annealed for two and three days shared a similar performance, and both of them performed better than the membrane annealed for one day. The flux of the commercial membrane (GVHP) is illustrated here for comparison. A2 and A3, having a superb average flux of 35 LMH in 10 h of operation, were 75% higher than GVHP (20 LMH).

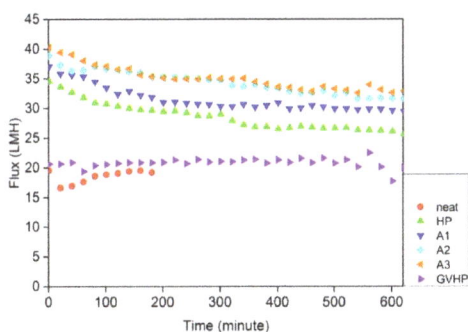

Figure 2. Comparison of flux performance between post-treated and as-spun electrospun membranes and commercial ones.

3.2. Increased Crystallinity and Appearance of α Phase after Annealing

FT-IR spectra of the α and β phases of PVDF in the PVDF-HFP copolymer were investigated comprehensively. It was found that the bands in terms of the β phase of PVDF appeared at 840 cm^{-1} and 1278 cm^{-1} [29,30] in all the as-spun and thermal-treated membranes, as shown in Figure 3; the bands related to the α phase of PVDF appeared at 615 cm^{-1}, 765 cm^{-1}, 795 cm^{-1}, 975 cm^{-1} and 1212 cm^{-1} [29–31], and they were only found in the annealed membranes. The phase of PVDF in the neat membrane was basically β, as the peak representing the α phase rarely appeared. Two factors contributed to the as-spun membranes mainly consisting of the β phase: (1) the existence of the HFP copolymer in the polymer chains [29]; (2) stretching and pulling of the electrospun fiber during the whipping process in electrospinning [31,32]. After the neat membrane was heat-pressed, the band at 840 cm^{-1} representing the β phase increased from 53% to 63.6%, while the band in terms of the α phase did not appear after the heat-press treatment. The increase in the transmittance of β phase bands indicated the increase in the crystallinity of the membrane as the amorphous phase of the PVDF was converted into the β phase due to the mechanical deformation, caused by the pressure applied on the membranes during heat-pressing [25,26]. Vineet et al. found that annealing PVDF above 80° led to an increase in both the total crystallinity and the α phase PVDF percentage [29,33]. Du et al. also pointed out that annealing of the membrane resulted in the transformation of some regions of the PVDF phase from β to α [29]. In this study, bands in terms of the α phase started to appear at 615 cm^{-1}, 765 cm^{-1}, and 975 cm^{-1} after one day of annealing (A1), and were more obvious at 615 cm^{-1}, 765 cm^{-1}, 795 cm^{-1}, 975 cm^{-1} and 1212 cm^{-1} after two and three days of annealing (A2 and A3), while the transmittance at the 840 cm^{-1} band decreased back to 62.1% which is same value as in the neat membrane. It is worth noting that the membrane annealed for three days (A3) had nearly identical FT-IR spectra results as the one annealed for two days (A2), which means two days of annealing could be long enough for sufficient state conversion. The appearance of the α phase PVDF in the membrane is favorable for the MD process owing to its non-polar properties because it leads to a decrease in the dipolar interaction between the water molecules and the membrane [34]. The existence of the α phase PVDF can increase the liquid entry pressure and, hence, the wetting resistance, contributing to better long-term MD performance. Saffarini et al. stated that annealing could also affect the microstructure evolution [35], which benefits the long-term MD operation as well. By increasing the crystallinity and releasing the internal stresses caused by the heat-pressing, the thermal stability of the membrane could be improved, preventing the LEP from dropping rapidly owing to distortion at a high feed temperature.

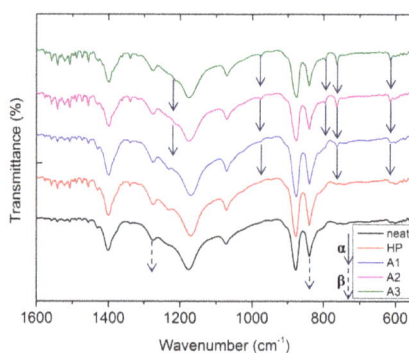

Figure 3. FT-IR (Fourier-transform infrared) spectra of as-spun and thermal-treated electrospun membranes.

3.3. Further Detection of Phase Conversion by XRD

To further confirm the change of the PVDF crystal phase in the PVDF-HFP membranes, X-ray diffraction was conducted on these membrane samples. Figure 4 shows the amorphous and both the α and β crystalline phases of PVDF in the PVDF-HFP electrospun membranes with or without post-treatment. It can be clearly seen that the as-spun electrospun membrane has many more regions containing the amorphous phase of PVDF as the peaks representing the α and β phases have been merged, leading to a much broader and weaker peak. It is worth noting that the amorphous state of the PVDF cannot be detected by FT-IR as discussed in last section. After heat treatment, the regions of the amorphous-state PVDF reduced greatly as two sharp peaks at 18.4° and 20.8° appeared, representing α (020) and β (110), respectively [29,30]. The sharp increase in the crystallinity of the PVDF-HFP membrane resulted in a great change of the membrane properties, especially the contact angle and LEP, which will be discussed in later sections. Annealing of the heat-pressed membrane can further improve the crystalline structure of the PVDF-HFP membrane in a favorable way. Due to the non-polar structure, the membrane composed of more regions of the α-phase PVDF had higher hydrophobicity, leading to a higher wetting resistance. While the total crystallinity maintained was nearly unchanged, annealing the heat-pressed membrane for one day at 120 °C could convert the β phase to the α phase as the relative peaks varied. After the membranes were further annealed for another day, both the α phase and crystallinity were increased, leading to the increase in the thermal and mechanical stability [30]. More interestingly, a sharp new peak with a high intensity at 28.1° appeared, which is very close to α (111) having a two-theta angle of 27.8°. Considering the various new peaks representing the α phase PVDF crystal appearing in the FT-IR in the A2 membranes, we assumed that the peak at the 2θ angle of 28.1° was α (111) [30,32]. The high intensity of α (111) means a big increase in the α phase PVDF as well as the total crystallinity, so we believe that more amorphous phase of the PVDF along with the β phase was converted to the α phase. A3 (the membrane annealed for three days) shows identical spectra of XRD results as A2, so it is not shown in Figure 4.

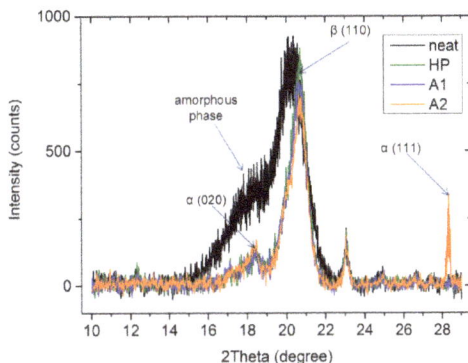

Figure 4. XRD (X-ray diffraction) patterns (2θ spectra) of as-spun and thermal-treated electrospun membranes.

3.4. Improvement of Membrane Properties

3.4.1. Pore Size Distribution and Porosity

Figure 5 shows the pore size distribution of membranes before and after the heat treatment. The as-spun membrane has an average pore diameter of 1.3 μm and a much wider range (from 0.8 μm to 2.2 μm) than the other samples. The heat-pressed membrane (HP) has the smallest average pore size (1.1 μm) and a much smaller range of the pore diameter compared with the neat membranes because the fibers were deformed and widened under high pressure at 150 °C [24]. Annealing the HP membranes for one day can further narrowed the pore size distribution, although the mean pore diameter increased slightly from 1.1 μm to 1.2 μm. The increase in the average pore size may be due to the further decrease in the membrane thickness from 39 μm to 37 μm, which might be caused by the release of internal stress generated during heat-pressing; the other possible cause could be the supplementary merging of the crystalline phase of the PVDF which was converted from the amorphous phase during annealing [30]. The results were consistent with the study of Saffarini et al., as it was found that an increase in the annealing temperature would lead to an increase in the average pore size of the membranes [35]. Further annealing for another day led to a greater decrease in the range of the pore diameter and a sharp peak appeared, although the average mean pore size was further slightly increased to 1.3 μm (which was same as that of the neat membrane). An increased average pore diameter led to an increase in the flux performance [36], justifying that MD applied with the A2 and A3 membranes had the highest flux performance compared to the other samples. In the literature, the pore size distribution of commercial MD membranes generally lies between 0.2 μm and 1 μm as the increase in the maximum pore size greatly decreases the LEP, leading to serious wetting problems [3,6]. Narrowing the range of the pore diameter greatly improves the wetting resistance as the LEPs of A2 and A3 were much higher than that of the neat membrane. The other factor regarding the improvement of the wetting resistance was the evolution of the crystalline structure in the micro-view, which was fully discussed in last two sections. The heat-pressing process turned the amorphous state of the PVDF into crystalline phases, which had much higher thermal and mechanical stability [30]; the annealing process helped some of the β- and amorphous-phase PVDF get converted into the α phase, which had higher hydrophobicity owing to its non-polar structure. Both factors contributed to A2's high wetting resistance in the MD process even though it had the highest average pore size (but a narrower pore size distribution). A high average pore size led to a high flux performance, improving the permeation efficiency of the MD. A3 had a nearly identical pore diameter distribution as A2; therefore, annealing for two days was sufficient for the improvement of the pore diameter distributions.

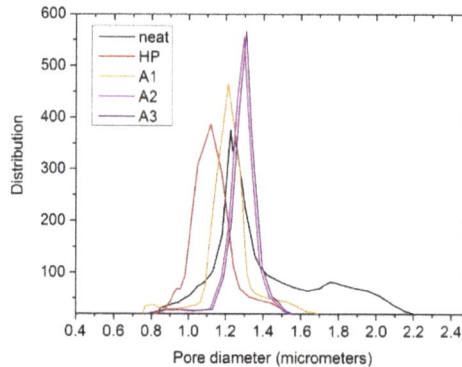

Figure 5. Pore diameter distribution of as-spun and post-treated electrospun membranes.

The porosities of the neat and thermal-treated membranes did not have significant differences. The porosities of the neat, HP, A1, A2, A3 membranes were 91%, 86%, 86%, 87% and 87%, respectively, somehow following the trend of the contact angle. However, their differences were too minimal to affect the MD flux performance.

3.4.2. LEP and Contact Angle

A comparison of the membrane properties of the LEP, contact angle and pore size between the as-spun and post-treated membranes was investigated. Consistent with the previous study [24], Figure 6 shows that the LEP of the neat membrane was the lowest, contributing to the rapid wetting in DCMD. After the membrane was heat-pressed, the LEP increased dramatically from 73 kPa to 91 kPa owing to the vastly decreased average pore diameter and increased crystallinity, although the contact angle decreased from 153° to 142° owing to the reduced roughness [24]. Annealing the HP membranes does not result in a significant change of the LEP as the effect of the increased pore sizes is offset by the increase in the hydrophobicity of the membranes, indicated by the increased contact angle. A2 shares a nearly identical LEP as the HP while it had the highest pore size (1.3 μm) owing to a greatly increased contact angle of 147°. Two days of annealing is proved to be sufficient for the improvement of the properties as A3 shares a nearly identical LEP, contact angle and average pore diameter as A2. It is worth mentioning that support layers had minimal effects on the LEP owing to the hydrophilicity and large pore size. The introduction of support layers into the DCMD test mainly aimed to improve the mechanical strength of the electrospun membranes. However, the use of the support layer increased the mass transfer resistance and caused additional temperature polarization, which requires further study [6].

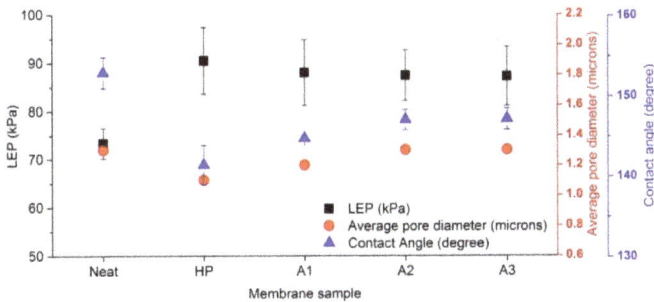

Figure 6. Comparison of LEP (liquid entry pressure), contact angle and average pore diameter of as-spun, heat-pressed, and annealed membranes.

4. Conclusions

In the present study, we successfully fabricated electrospun nanofiber membranes and they were modified by post-treatments such as heat-pressing and annealing. Especially, the effects of annealing on the improvement of the MD flux performance were investigated. Based on the results, it was found that the sufficiently annealed electrospun membranes had a slightly increased average pore size, which contributed to their much higher flux in the DCMD compared to the other membrane samples. An average flux of 35 LMH could be achieved for annealed membranes while an average flux of 20 LMH was obtained by using commercial PVDF membranes in the same DCMD setup during 10 h of operation. It was found that the increase in the α phase of PVDF led to the increase in the hydrophobicity of the membrane, so the high LEP after heat-pressing remained unchanged, although the average pore diameter increased. The other key factor in terms of the wetting resistance was the greatly narrowed pore size distribution as the membranes annealed for two and three days had the narrowest distribution among all the samples. Therefore, the higher average pore size of the annealed electrospun membranes contributed to the improvement of the flux performance while its negative effect on the LEP was offset by the increase in the hydrophobicity, preventing wetting issues. In summary, along with heat-pressing, the annealing technique is recommended for electrospun nanofiber membranes to improve their characteristics such as LEP, contact angle and pore size, which is key for a better permeation and rejection performance of MD for desalination. The use of nanofiber membranes with proper tuning of the material, process and post-treatment parameters to obtain near-ideal MD membranes with high performance efficiency can potentially accelerate their industrial applications.

Acknowledgments: This research was supported by a grant (16IFIP-B065893-04) from the Industrial Facilities & Infrastructure Research Program funded by the Ministry of Land, Infrastructure and Transport of the Korean government. The authors also acknowledge the grant from the ARC Future Fellowship (FT140101208) and the UTS FEIT Blue Sky Research Fund.

Author Contributions: M.Y., L.D.T. and H.K.S. conceived and designed the experiments; M.Y., Y.C.W. and C.C. performed the experiments, helped with data acquisition and prepared the figures; M.Y., Y.C.W., L.D.T. and H.K.S. analyzed the data; M.Y., Y.C.W. and L.D.T. wrote the paper; and H.K.S. and L.D.T. supervised and approved the research plan.

Conflicts of Interest: The authors declare no conflict of interest.

References

1. Efome, J.E.; Rana, D.; Matsuura, T.; Lan, C.Q. Enhanced performance of pvdf nanocomposite membrane by nanofiber coating: A membrane for sustainable desalination through md. *Water Res.* **2016**, *89*, 39–49. [CrossRef] [PubMed]

2. Drioli, E.; Ali, A.; Macedonio, F. Membrane distillation: Recent developments and perspectives. *Desalination* **2015**, *356*, 56–84. [CrossRef]
3. Camacho, L.M.; Dumée, L.; Zhang, J.; Li, J.-D.; Duke, M.; Gomez, J.; Gray, S. Advances in membrane distillation for water desalination and purification applications. *Water* **2013**, *5*, 94–196. [CrossRef]
4. Cho, D.-W.; Song, H.; Yoon, K.; Kim, S.; Han, J.; Cho, J. Treatment of simulated coalbed methane produced water using direct contact membrane distillation. *Water* **2016**, *8*, 194. [CrossRef]
5. Alkhudhiri, A.; Darwish, N.; Hilal, N. Membrane distillation: A comprehensive review. *Desalination* **2012**, *287*, 2–18. [CrossRef]
6. Eykens, L.; de Sitter, K.; Dotremont, C.; Pinoy, L.; van der Bruggen, B. How to optimize the membrane properties for membrane distillation: A review. *Ind. Eng. Chem. Res.* **2016**, *55*, 9333–9343. [CrossRef]
7. Guillen-Burrieza, E.; Mavukkandy, M.O.; Bilad, M.R.; Arafat, H.A. Understanding wetting phenomena in membrane distillation and how operational parameters can affect it. *J. Membr. Sci.* **2016**, *515*, 163–174. [CrossRef]
8. Ke, H.; Feldman, E.; Guzman, P.; Cole, J.; Wei, Q.; Chu, B.; Alkhudhiri, A.; Alrasheed, R.; Hsiao, B.S. Electrospun polystyrene nanofibrous membranes for direct contact membrane distillation. *J. Membr. Sci.* **2016**, *515*, 86–97. [CrossRef]
9. Woo, Y.C.; Tijing, L.D.; Park, M.J.; Yao, M.; Choi, J.-S.; Lee, S.; Kim, S.-H.; An, K.-J.; Shon, H.K. Electrospun dual-layer nonwoven membrane for desalination by air gap membrane distillation. *Desalination* **2017**, *403*, 187–198. [CrossRef]
10. Ma, M.; Hill, R.M.; Lowery, J.L.; Fridrikh, S.V.; Rutledge, G.C. Electrospun poly(styrene-block-dimethylsiloxane) block copolymer fibers exhibiting superhydrophobicity. *Langmuir* **2005**, *21*, 5549–5554. [CrossRef] [PubMed]
11. Zhang, J.; Li, J.-D.; Gray, S. Effect of applied pressure on performance of ptfe membrane in dcmd. *J. Membr. Sci.* **2011**, *369*, 514–525. [CrossRef]
12. Woo, Y.C.; Kim, Y.; Shim, W.-G.; Tijing, L.D.; Yao, M.; Nghiem, L.D.; Choi, J.-S.; Kim, S.-H.; Shon, H.K. Graphene/PVDF flat-sheet membrane for the treatment of ro brine from coal seam gas produced water by air gap membrane distillation. *J. Membr. Sci.* **2016**, *513*, 74–84. [CrossRef]
13. Prince, J.; Rana, D.; Singh, G.; Matsuura, T.; Kai, T.J.; Shanmugasundaram, T. Effect of hydrophobic surface modifying macromolecules on differently produced pvdf membranes for direct contact membrane distillation. *Chem. Eng. J.* **2014**, *242*, 387–396. [CrossRef]
14. Hwang, H.J.; He, K.; Gray, S.; Zhang, J.; Moon, I.S. Direct contact membrane distillation (DCMD): Experimental study on the commercial ptfe membrane and modeling. *J. Membr. Sci.* **2011**, *371*, 90–98. [CrossRef]
15. Tijing, L.D.; Choi, J.-S.; Lee, S.; Kim, S.-H.; Shon, H.K. Recent progress of membrane distillation using electrospun nanofibrous membrane. *J. Membr. Sci.* **2014**, *453*, 435–462. [CrossRef]
16. Liao, Y.; Wang, R.; Fane, A.G. Fabrication of bioinspired composite nanofiber membranes with robust superhydrophobicity for direct contact membrane distillation. *Environ. Sci. Technol.* **2014**, *48*, 6335–6341. [CrossRef] [PubMed]
17. Feng, C.; Khulbe, K.C.; Matsuura, T.; Tabe, S.; Ismail, A.F. Preparation and characterization of electro-spun nanofiber membranes and their possible applications in water treatment. *Sep. Purif. Technol.* **2013**, *102*, 118–135. [CrossRef]
18. Ahmed, F.E.; Lalia, B.S.; Hashaikeh, R. A review on electrospinning for membrane fabrication: Challenges and applications. *Desalination* **2015**, *356*, 15–30. [CrossRef]
19. Prince, J.; Rana, D.; Matsuura, T.; Ayyanar, N.; Shanmugasundaram, T.; Singh, G. Nanofiber based triple layer hydro-philic/-phobic membrane-a solution for pore wetting in membrane distillation. *Sci. Rep.* **2014**, *4*, 6949. [CrossRef] [PubMed]
20. Lee, E.-J.; An, A.K.; He, T.; Woo, Y.C.; Shon, H.K. Electrospun nanofiber membranes incorporating fluorosilane-coated TiO_2 nanocomposite for direct contact membrane distillation. *J. Membr. Sci.* **2016**, *520*, 145–154. [CrossRef]
21. Tijing, L.D.; Woo, Y.C.; Shim, W.-G.; He, T.; Choi, J.-S.; Kim, S.-H.; Shon, H.K. Superhydrophobic nanofiber membrane containing carbon nanotubes for high-performance direct contact membrane distillation. *J. Membr. Sci.* **2016**, *502*, 158–170. [CrossRef]

22. Ren, L.-F.; Xia, F.; Shao, J.; Zhang, X.; Li, J. Experimental investigation of the effect of electrospinning parameters on properties of superhydrophobic PDMS/PMMA membrane and its application in membrane distillation. *Desalination* **2017**, *404*, 155–166. [CrossRef]

23. An, A.K.; Guo, J.; Lee, E.-J.; Jeong, S.; Zhao, Y.; Wang, Z.; Leiknes, T. PDMS/PVDF hybrid electrospun membrane with superhydrophobic property and drop impact dynamics for dyeing wastewater treatment using membrane distillation. *J. Membr. Sci.* **2017**, *525*, 57–67. [CrossRef]

24. Yao, M.; Woo, Y.C.; Tijing, L.D.; Shim, W.-G.; Choi, J.-S.; Kim, S.-H.; Shon, H.K. Effect of heat-press conditions on electrospun membranes for desalination by direct contact membrane distillation. *Desalination* **2016**, *378*, 80–91. [CrossRef]

25. Na, H.; Zhao, Y.; Zhao, C.; Zhao, C.; Yuan, X. Effect of hot-press on electrospun poly(vinylidene fluoride) membranes. *Polym. Eng. Sci.* **2008**, *48*, 934–940. [CrossRef]

26. Tijing, L.D.; Woo, Y.C.; Johir, M.A.H.; Choi, J.-S.; Shon, H.K. A novel dual-layer bicomponent electrospun nanofibrous membrane for desalination by direct contact membrane distillation. *Chem. Eng. J.* **2014**, *256*, 155–159. [CrossRef]

27. Liao, Y.; Wang, R.; Tian, M.; Qiu, C.; Fane, A.G. Fabrication of polyvinylidene fluoride (PVDF) nanofiber membranes by electro-spinning for direct contact membrane distillation. *J. Membr. Sci.* **2013**, *425–426*, 30–39. [CrossRef]

28. Lalia, B.S.; Guillen-Burrieza, E.; Arafat, H.A.; Hashaikeh, R. Fabrication and characterization of polyvinylidenefluoride-*co*-hexafluoropropylene (PVDF-HFP) electrospun membranes for direct contact membrane distillation. *J. Membr. Sci.* **2013**, *428*, 104–115. [CrossRef]

29. Du, C.-H.; Zhu, B.-K.; Xu, Y.-Y. The effects of quenching on the phase structure of vinylidene fluoride segments in PVDF-HFP copolymer and PVDF-HFP/PMMA blends. *J. Mater. Sci.* **2006**, *41*, 417–421. [CrossRef]

30. Liu, J.; Lu, X.; Wu, C. Effect of preparation methods on crystallization behavior and tensile strength of poly(vinylidene fluoride) membranes. *Membranes* **2013**, *3*, 389–405. [CrossRef] [PubMed]

31. Salimi, A.; Yousefi, A. Analysis method: Ftir studies of β-phase crystal formation in stretched PVDF films. *Polym. Test.* **2003**, *22*, 699–704. [CrossRef]

32. Esterly, D.M. Manufacturing of Poly(vinylidene fluoride) and Evaluation of Its Mechanical Properties. Master's Thesis, Virginia Polytechnic Institute and State University, Blacksburg, CA, USA, 2002.

33. Tiwari, V.; Srivastava, G. Effect of thermal processing conditions on the structure and dielectric properties of pvdf films. *J. Polym. Res.* **2014**, *21*, 1–8. [CrossRef]

34. Aqeel, S.M.; Wang, Z.; Than, L.; Sreenivasulu, G.; Zeng, X. Poly(vinylidene fluoride)/poly(acrylonitrile)-based superior hydrophobic piezoelectric solid derived by aligned carbon nanotubes in electrospinning: Fabrication, phase conversion and surface energy. *RSC Adv.* **2015**, *5*, 76383–76391. [CrossRef] [PubMed]

35. Saffarini, R.B.; Mansoor, B.; Thomas, R.; Arafat, H.A. Effect of temperature-dependent microstructure evolution on pore wetting in ptfe membranes under membrane distillation conditions. *J. Membr. Sci.* **2013**, *429*, 282–294.

36. Manawi, Y.M.; Khraisheh, M.; Fard, A.K.; Benyahia, F.; Adham, S. Effect of operational parameters on distillate flux in direct contact membrane distillation (DCMD): Comparison between experimental and model predicted performance. *Desalination* **2014**, *336*, 110–120. [CrossRef]

applied
sciences

MDPI

Article

Improving Liquid Entry Pressure of Polyvinylidene Fluoride (PVDF) Membranes by Exploiting the Role of Fabrication Parameters in Vapor-Induced Phase Separation VIPS and Non-Solvent-Induced Phase Separation (NIPS) Processes

Faisal Abdulla AlMarzooqi, Muhammad Roil Bilad and Hassan Ali Arafat *

Department of Chemical and Environmental Engineering, Masdar Institute of Science and Technology,
P.O. Box 54224 Abu Dhabi, UAE; falmarzooqi@masdar.ac.ae (F.A.A.); roilbilad130@yahoo.com (M.R.B.)
* Correspondence: harafat@masdar.ac.ae; Tel.: +971-2-810-9119

Academic Editor: Enrico Drioli
Received: 25 December 2016; Accepted: 4 February 2017; Published: 14 February 2017

Abstract: Polyvinylidene fluoride (PVDF) is a popular polymer material for making membranes for several applications, including membrane distillation (MD), via the phase inversion process. Non-solvent-induced phase separation (NIPS) and vapor-induced phase separation (VIPS) are applied to achieve a porous PVDF membrane with low mass-transfer resistance and high contact angle (hydrophobicity). In this work, firstly, the impacts of several preparation parameters on membrane properties using VIPS and NIPS were studied. Then, the performance of the selected membrane was assessed in a lab-scale direct-contact MD (DCMD) unit. The parametric study shows that decreasing PVDF concentration while increasing both relative humidity (RH) and exposure time increased the contact angle and bubble-point pore size (BP). Those trends were investigated further by varying the casting thickness. At higher casting thicknesses and longer exposure time (up to 7.5 min), contact angle (CA) increased but BP significantly decreased. The latter showed a dominant trend leading to liquid entry pressure (LEP) increase with thickness.

Keywords: membrane distillation; polyvinylidene fluoride; hydrophobic; non-solvent-induced phase separation; vapor-induced phase separation

---

1. Introduction

The membrane distillation (MD) process is driven by the vapor pressure difference between a hot and a cold stream in the feed and permeate side of a membrane module, respectively. This driving force is created by the temperature difference between the hot and cold streams. Both streams are separated by a membrane that ideally features the following properties: high hydrophobicity, high porosity, low tortuosity, and low thickness. In addition to low mass transfer resistance, an MD membrane should be able to withstand a certain liquid entry pressure (LEP) to avoid wetting [1]. Theoretical LEP of a membrane is a function of maximum pore size (also called bubble point, BP) and surface hydrophobicity, expressed in terms of the surface contact angle (CA) with water. LEP can be estimated using the Cantor–Laplace equation [1].

Polyvinylidene fluoride (PVDF) is an attractive MD membrane material, which can be made into membranes via phase inversion. It has low surface energy, good thermal stability and low conductivity. In the phase inversion process, a thermodynamically-stable polymer solution (mixture of polymer and its solvent) is brought to instability by several means: non-solvent-induced phase separation (NIPS), vapor-induced phase separation (VIPS) or temperature-induced phase separation (TIPS). All of these have been applied to prepare PVDF-based MD membranes [2–4]. The NIPS process is the most popular and has been used to prepare different types of membranes, including reverse osmosis, nanofiltration, ultrafiltration, microfiltration and MD [5,6]. When used to produce PVDF membranes in a strong solvent/non-solvent system, typically, the NIPS process is characterized by a relatively fast mass transfer rate (instantaneous demixing, relative to VIPS) leading to asymmetric structure (dense top layer, supported by a more porous layer) [5], narrow pore size distribution (PSD), relatively low mean flow pore size (MPS) and maximum pore size (BP) very close to MPS; all are desired properties of an MD membrane. However, under those same conditions, NIPS also produces a membrane with a fairly smooth surface (low roughness) leading to a low surface energy (high CA), which promotes low LEP and thus membrane wetting.

Increasing the CA of PVDF membranes has been demonstrated by applying weak non-solvent (ethanol) in a single bath system [7] and weak non-solvent (ethanol) as the first bath followed by strong non-solvent (water) in the second bath of a dual-bath coagulation system [8]. These methods increased the CA significantly (up to 153°) but the resulting membranes had low porosity originating from a dense structure as a result of a delayed demixing that led to a high mass-transfer resistance. Superhydrophobic PVDF membranes can also be obtained via the VIPS process, but feature large BP that diminishes the LEP [8]. A study which investigated the phase inversion mechanisms that led to the formation of a highly hydrophobic surface [3] emphasized the importance of solution composition pathway when crossing both binodal and spinodal lines in the three-phase diagram. Depending on the route of the composition changes, porous and spherulitic structures resulting from PVDF crystallization resulted in a hydrophobic surface, while net-like morphology as a result of spinodal-decomposition (SD) resulted in a superhydrophobic surface [8]. The latter can be produced under low PVDF concentrations (<5%), high relative humidity (RH) and high vapor temperature. However, too-low PVDF concentrations produce membranes that are vulnerable to wetting and low-salt rejections (SR). At such low concentrations, there is a lack of mechanical stability in the membrane and the anchoring of the membrane on the support layer becomes problematic. To avoid this problem, Fan et al. [9] applied the co-casting technique, consisting of high (16% *w/w*) and low (10% *w/w*) PVDF concentrations at the bottom and top layer, respectively, to achieve a highly hydrophobic surface by ensuring low PVDF concentration on the top of the film.

In this study, the interplays between surface and bulk membrane properties on one hand and the fabrication parameters during phase inversion, on the other hand, are disentangled by systematically exploiting the role of VIPS and NIPS parameter variations on the membrane characteristic. This study is distinguished from previous studies in trying to relate changes in surface properties to parameters in the membrane fabrication process, through a systematic disentanglement of the latter. It utilizes the fact that the water vapor absorption/imbibition rates are much faster than water/solvent diffusion rates, resulting in a gradient of composition profile across the casting film thickness (being the lowest at the top of film) during film exposure to humid atmosphere [10]. The surface structure mixture of nano- and micro-scale roughness [7], which promotes higher hydrophobicity, is expected. Meanwhile, this surface structure maintains a pore structure that controls salt rejection. Membrane formation mechanism via VIPS was first studied, followed by a series of tests to investigate the impact of polymer concentration, relative humidity (RH) and exposure time. The role of thickness was then investigated, followed by a moisture uptake test as a way to experimentally support our findings. Finally, the performance of the selected membrane was assessed in a lab-scale direct-contact MD (DCMD) unit.

2. Materials and Methods

2.1. Membrane Preparation

All membranes were prepared using PVDF polymer (HSV900, Mw 92,840 kDa, Arkema, Colombes, France), dimethylacetamide (DMAC, Sigma-Aldrich, St. Louis, NA, USA) and deionized (DI) water as the polymer, solvent and non-solvent, respectively. In the NIPS process, after being thoroughly dissolved and degassed, the polymer solution was cast on a non-woven support (NWS) (Novatexx 2471, donated by Freudenberg-filter, Weinheim, Germany) at 24 °C using a doctor blade with adjustable height to give a wet-casting thickness of 250, 500, 750 or 1000 µm (henceforth referred to as the "casting thickness"). This was immediately followed by immersion in a coagulation bath containing deionized water. A number of membrane samples were prepared via a VIPS process which was similar to the NIPS process except that it involved film exposure to humid air at different percentages of relative humidity (RH): 37%, 60% and 80%. This exposure was after the casting step, for a certain time (henceforth called "exposure time") followed by immersion into a coagulation bath (DI water at room temperature). The details of preparation parameters for each sample are shown in Table 1.

Table 1. Summary of membrane preparation parameters.

Parameter	Range
Polyvinylidene fluoride (PVDF) concentration	8%, 10%, 12% and 15% (*w*/*w*)
Relative humidity	N/A *, 37% (room relative humidity), 60% and 80%
Exposure time	0 *, 2, 5, 10, 30, 60 and 120 min
Casting thickness	250, 500, 750 and 1000 µm

* Not Applicable: For non-solvent-induced phase separation (NIPS) process.

2.2. Membrane Characterization

The microstructure of the membrane surface was observed using scanning electron microscopy (SEM, Quanta-250, FEI, Hillsboro, OR, USA). Samples were gold and palladium-coated at 50 Å thickness prior to SEM analysis. The MPS, BP and PSD were measured using a capillary flow porometer (CFP, Porous Materials Inc., Ithaca, NY, USA). Volume porosity was calculated experimentally using a gravimetric method using Galwick® as a wetting liquid. The CA of DI water on the membrane and NWS surfaces was measured using the sessile drop method with a contact angle goniometer (Krüss DSA 10 Mk2, Hamburg, Germany) at 24 °C. Multiple measurements (at least six) at different locations of the same membrane sample were taken to enhance data accuracy.

2.3. Moisture Uptake Test

The cast film on the NWS was placed on a digital mass balance in a controlled humidity chamber. The evolution of weight was measured and recorded over time. The tests were performed for the three applied RHs, 37%, 60% and 80%. This test works on the assumption that the evaporation rate of DMAC is very low, i.e., negligible compared to the water vapor imbibition rates into the cast film. This assumption is based on the fact that there is an overall increase in the membrane weight with time during the water vapor imbibition stage. It is also worth noting that water is much more volatile than DMAC (water and DMAC vapor pressures are 3167 Pa and 557 Pa, respectively, at 25 °C) [11,12].

2.4. DCMD Experiments

DCMD experiments were carried out under controlled conditions using a custom built lab-scale DCMD setup. The description of the set-up and its operation can be found elsewhere [13]. The feed- and permeate-side liquids were 35 g/L NaCl solution and DI water, respectively. Both were circulated at a constant flow rate of 25 L/h (corresponding to a linear flow velocity of 0.08 m/s in the DCMD chamber). During each experiment (~3 h), the feed and permeate temperatures were maintained at 70 and 25 °C, respectively. Permeate overflow was continuously measured by means of a microbalance and the membrane flux was calculated from this value.

The salt rejection (SR) was calculated based on the following formula:

$$SR\ (\%) = \left(1 - \frac{C_{Permeate}}{C_{Feed}}\right) \times 100$$

where C_{feed} and $C_{permeate}$ are the salt concentrations in the feed and permeate streams, respectively (g/L).

3. Results and Discussion

3.1. Membrane Morphologies in NIPS vs. VIPS

Two distinct membrane morphologies were obtained for membranes prepared using NIPS and VIPS processes. In general, VIPS membranes, irrespective of applied RH, show a much higher CA (>125°) than the NIPS membrane (CA = 90.74°), as shown in Table 2. This finding regarding VIPS morphology was also reported elsewhere [14]. When calculating the LEP of membranes prepared using both processes, they all range from 103 to 107 kPa, which makes the membranes vulnerable to wetting. This low LEP range, mostly caused by the high BP values for all membranes, disallows the application of high operational pressure (typically up to 200 kPa depending on MD module system and desired cross-flow velocity) that is required to overcome pressure-drop in both the feed and the permeate streams in a module system. Additional pressure is also required to induce sufficient cross-flow velocity for controlling concentration polarization on the membrane surface. The advantage of high CA from the VIPS and low BP from NIPS to achieve high LEP is diminished because of large BP (10–21 μm) of VIPS membranes and low CA of NIPS membranes. Nevertheless, this finding brings about a potential for disentangling these features in the two processes to gain the advantages of both processes (high CA and low BP). As mentioned at the bottom of Table 2, all the LEP values in this table and throughout this manuscript were calculated using the Cantor–Laplace equation,

$$LEP = \frac{-2\gamma \cos\theta}{r_{max}}$$

where γ is the surface tension of the wetting liquid (in this case water at 25 °C, 0.07199 N/m), θ is the contact angle between the membrane and the wetting liquid (water) and r_{max} is the maximum pore radius of the membrane (i.e., bubble-point pore size). It is worth noting that the assumption in this equation is that pores are perfectly cylindrical with a constant radius of curvature.

Table 2. Summary of membrane properties prepared using NIPS and vapor-induced phase separation (VIPS) process, prepared under casting thickness of 500 μm using 12% w/w PVDF concentration.

Membrane	MPS (μm)	BP (μm)	CA	LEP (kPa) *
NIPS	0.08	0.38	90.7	103.91
VIPS (37% RH)	7.80	12.25	136.3	106.79
(60% RH)	5.85	21.19	130.9	103.56
(80% RH)	6.07	10.63	127.2	106.55

* Calculated using Cantor–Laplace equation. MPS: mean flow pore size; BP: bubble-point pore size; CA: contact angle; LEP: liquid entry pressure; RH: relative humidity.

Since all membranes were prepared from the same material, the difference in the surface morphology can thus be held responsible for the change in surface CA values, as can be seen in Table 2 and Figure 1. It is known that highly hydrophobic surface is a result of the presence of multi-scale surface features consisting of both micro- and nano-scale roughness [7,15–17], also known as the lotus leaf effect. The high contact angle on the surface of a lotus leaf is a consequence of such micro- and nano-scale roughness, the micro- and nano-scale roughness can be clearly seen in our case for VIPS-60% and VIPS-80% SEM images, Figure 1. The micron-sized pore mouth most likely contributed to the micro-scale roughness for both VIPS and NIPS membranes. The higher CA of the VIPS membranes originated from the nano-scale roughness from nodule-like morphology (circled features in Figure 1) that most probably originated from small spherulites of PVDF crystals. These can be clearly seen in the VIPS-60% and VIPS-80% SEM images.

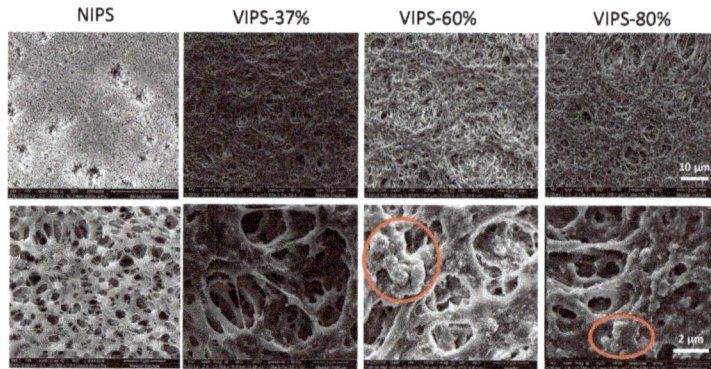

Figure 1. Top surface scanning electron microscope (SEM) images of membranes prepared using NIPS (non-solvent-induced phase separation) and VIPS (vapor-induced phase separation) processes with 12% *w/w* PVDF (Polyvinylidene fluoride) concentration and cast thickness of 500 μm (10 μm and 2 μm scale bares for all top and bottom row SEM images, respectively).

The formation of more crystalline structure in the VIPS membranes was possible due to the nature of compositional change pathway during the phase inversion process. This pathway can be explained using a three-component system phase diagram (Figure 2). Starting with a NIPS process (path D), spontaneous demixing occurs immediately once the polymer solution gets in contact with the non-solvent in the coagulation bath [18–20]. In this case, the rate of non-solvent intrusion into the casted polymer film is much higher than the rate of solvent extrusion, leading to highly porous membrane as discussed in detail by Khayat and Matsuura (2011). As the compositional pathway moves from the initial polymer solution composition to the final membrane composition, it enters the two-phase region where it splits into polymer-rich (pint 1, Figure 2) and lean (point 2, Figure 2) phases connected by a tie-line. As the final membrane composition is approached, both solvent and non-solvent eventually evaporate during the drying period and the membrane becomes a single-phase polymer solid. At the final membrane composition, the two phases are at thermodynamic equilibrium. A polymer-rich phase with composition S is marked in Figure 2 as well as a polymer-lean phase with composition L. Point L represents the composition of the voids or pores of the membrane which contain water which eventually evaporate during the drying stage. At the other extreme, when a VIPS process is applied without the final coagulation bath step, delayed demixing leads the compositional pathway to follow path A (Figure 2) [21]. In this case, the rate of non-solvent intrusion into the casted polymer film is much slower than the rate of solvent extrusion. This path represents the case of lowest RH, where the composition of the polymer solution never crosses the binodal curve and maintains a single phase. As solvent and non-solvent exchange occurs at a rate much slower than in the NIPS case, there are three different mass-transfer processes occurring simultaneously; water vapor is absorbed into the polymer solution while DMAC and water evaporate, leaving the polymer solution. The former absorption rate of water is relatively much higher than the evaporation as will be discussed later (Section 3.3). The driving force for the water absorption in this case is the chemical potential difference between the atmosphere surrounding the polymer solution and the polymer solution itself. Over time, the difference between these two chemical potentials decreases, causing the rate of water vapor absorption to change as the membrane goes through the compositional pathway A. The decrease in chemical potential is depicted in the compositional pathway of A by the increasing slope of the path, representing a decreasing intake of water and a sharper movement towards the PVDF-rich phase. As the composition crosses the crystallization line, the polymer solution solidifies, forming crystal nuclei and eventually forming a membrane with crystalline structures (evidence of this is discussed later, Figure 6). Compositional paths B–F and C–E (Figure 2) represent a combination of a complete NIPS (pathway D) and a complete VIPS (pathway A) process [21,22]. As the RH increases, the compositional path moves from pathway A towards D, with B–F and C–E representing two RH cases with B–F having a lower RH than C–E. As the RH increases, the chemical potential difference between the polymer solution and the atmosphere surrounding it increases as well, leading to a higher rate of absorption of water (as can be seen later in Figure 8). Pathway B–F starts in a similar way to path A, but with a higher RH. As the polymer solution loses DMAC and gains water it enters the meta-stable region [21]. Once there, the casted film is immersed in the coagulation bath. At this point the pathway shifts from B to F, following a path similar to D but at a different initial point in the two-phase region. Ultimately, the membrane forms with different surface morphology, pore properties and porosity [18]. Pathway C is similar to B but with an even higher RH and a different exposure time. Here, the membrane is immersed in the coagulation bath after it crosses the spinodal line [21] moving from path C to E.

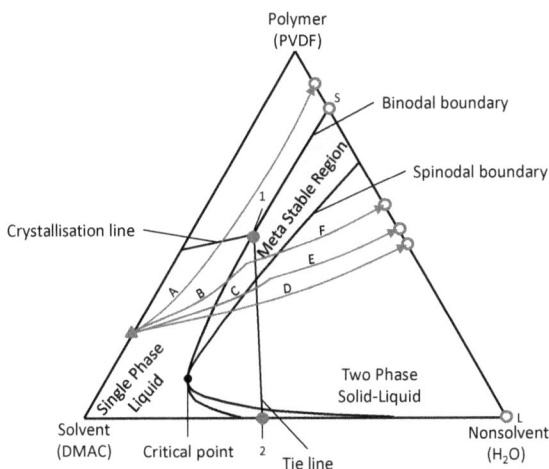

Figure 2. A three-component phase diagram for the polymer, solvent and non-solvent system generally depicting the VIPS and NIPS processes. Path A and D represent delayed (VIPS) and spontaneous (NIPS) de-mixing, respectively. Paths B and C represent an initial VIPS process followed by a NIPS. ▲ is initial polymer solution composition and ○ is composition of final membrane. DMAC: dimethylacetamide.

Unlike the NIPS process, in which introducing the cast film into the coagulation bath promotes instantaneous demixing (Figure 1), slow uptake of water from humid air allows the composition change to cross or approach the crystallization region in the phase diagram (Figure 2) and form crystalline morphology, allowing the crystal nuclei to grow thanks to the semi-crystalline property of PVDF, as observed in the case of VIPS-37%, VIPS-60% and VIPS-80% in Figure 1 [23]. This way, crystal nuclei were formed and they had sufficient time to grow and produce a spherulitic structure in the resulting membrane as illustrated in Figure 1. The large PVDF agglomerates are highly crystalline spherulites (micron size), while the nodule-like morphologies over the surface of the spherulites are the ends of the PVDF polymer chains. This pathway also alters the membrane pore size to be much larger, as shown in Figure 3. On one hand, the rapid demixing process disallows further growth of the polymer-lean phase, resulting in small pore size of NIPS membranes [24]. On the other hand, slow uptake of water from humid air gives sufficient time for the polymer-lean hase for growth and coalescence after the composition enters the meta-stable region and before phase inversion is completed, creating the large pores in the VIPS-made membranes.

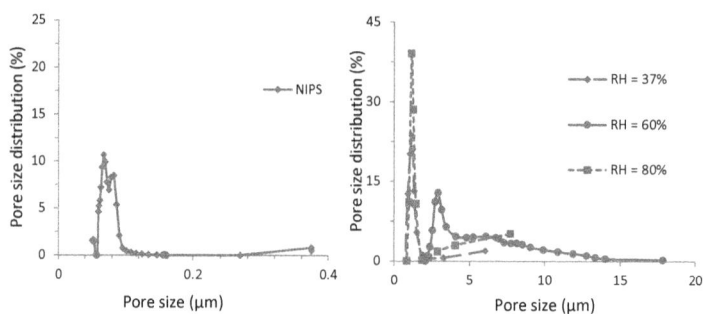

Figure 3. Pore size distributions (PSD) of membranes made via the NIPS (**left**) and VIPS (**right**) processes. Casting thickness: 500 μm, polymer solution: 12% w/w PVDF.

3.2. Going from NIPS to VIPS: A Parametric Study

As an attempt to combine the advantages of both VIPS and NIPS processes, a series of parametric studies to understand the role of several parameters on structure, CA and BP, were performed. The main objective is to obtain membranes which are highly porous with a surface morphology in favor of high CA, while keeping an acceptable BP (0.4–0.6 μm).

3.2.1. Effect of PVDF Concentration

Polymer concentration plays a significant role in determining the demixing process. Lower polymer concentrations are more likely to experience spinodal-decomposition (SD), producing a net-like surface structure in favor of the superhydrophobic property because their initial compositions are closer to the critical point in which binodal and spinodal lines are in a very close proximity [5]. Indeed, it was found that PVDF concentration affected membrane surface morphology, as shown in Figure 4. However, a net-like structure associated with superhydrophobicity originating from SD was not observed for all membrane samples prepared in this study. It was only observed in the 8% case as can be seen in the inset of the 8% membrane in Figure 4. This is mainly because of relatively high applied concentrations of PVDF compared to the one applied by [5]. Net-like structure is expected only during SD at very low PVDF concentrations and under high casting temperature (60°C) and high RH (100%). Nevertheless, a low degree of spherulitic morphology coupled with a large number of nano-scale nodules (Figure 4) was observed for 8% PVDF solution, leading to a high CA of 123° (Table 3).

As PVDF concentration increased, a decrease in both surface porosity as well as mean pore size was observed (Table 3). The change in PVDF concentrations clearly affected the hydrophobicity (CA) of the membranes. As the concentration increased, the CA decreased. This can be attributed to the flat morphology at a higher concentration which led to almost hydrophilic membranes (CA of 15% PVDF was below 90°).

The advantage of membranes prepared from lower PVDF concentrations is diminished by their large BP and MPS (Table 3). Therefore, the obtained LEP data suggests that a desirable PVDF concentration is around 12%. Membranes prepared using lower PVDF concentrations exhibit hydrophobic properties thanks to rough surface structure, but feature a rather large BP. On the other hand, the membrane made from a higher PVDF concentration (15%) had a very flat surface that reduced its CA, but offered a substantially low BP. As a result, by merely tailoring PVDF concentration, an optimum concentration can be ball-parked with respect to contact angle and pore sizes. In this case, the value is 12%.

Table 3. Properties of membranes cast using PVDF solutions of different concentrations at cast thickness: 500 μm, relative humidity (RH): 60% and exposure time: 30 min.

PVDF Concentration (*w/w*)	MPS (μm)	BP (μm)	CA (°)	LEP (kPa)	Volume Porosity
8%	0.2076	7.5956	123.2	108.30	48%
10%	0.1196	4.0325	113.1	111.21	70%
12%	0.089	0.210	94.6	143.93	88%
15%	0.064	0.652	88.9	96.64	67%

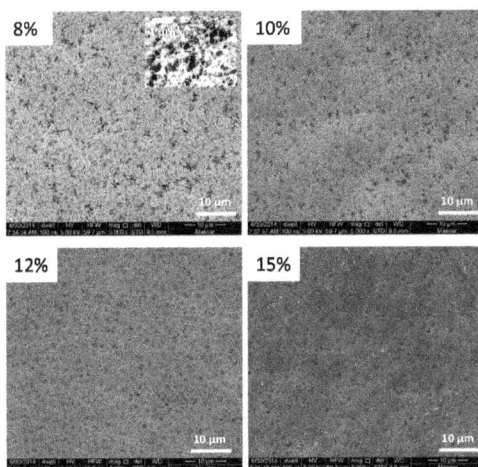

Figure 4. Surface SEM images of membrane samples prepared using polymer solutions of different PVDF concentrations (casting thickness: 500 μm, RH: 60%, exposure time: 30 min).

3.2.2. Effect of Humidity

Higher RH provides larger chemical potential for water to absorb on the cast film. Therefore, for a similar period of time, a higher amount of water uptake is expected for a higher RH (later discussed in Section 3.3). By considering the high miscibility of water and DMAC, the transport of DMAC from the deeper zone of the cast film to the top of the film and influx of water from the top of the film is expected, leading to lower concentration of PVDF at the top of the film. This is due to the transport of DMAC from deep in the bulk of the casted film to the top layer, leaving behind a PVDF-rich phase at the bottom layers of the cast film and a PVDF-lean phase at the top layers. Increasing humidity and/or exposure time thus promotes a higher concentration gradient of PVDF that promotes a change of surface hydrophilicity, like the one observed as an effect of PVDF concentration. This is in agreement with the obtained CA results seen in Table 4. CA increased from 91.0° to 104.5° and 93.6° to 98.2° by increasing exposure time from 2 to 10 min for RHs of 60% and 80%, respectively.

As the RH and exposure time increased, the presence of droplet-like pore mouths on the membrane surface increased (see SEM images of 80% at 10 min, Figure 5), which led to the creation of larger pores, especially at higher RH. These observations are in agreement with BP and MPS results, both of which increased with an increase in RH (Table 4). It is worth noting that some large pores on the membrane surface were a result of localized water vapor condensation on the top of the polymer film that was unavoidable sometimes during RH tests. Whenever the ambient humidity was below the RH set-point, the humidifier created extra water vapor that could potentially precipitate on the top of polymer film, leading to local concentration of water.

Table 4. Membranes properties with casting thickness of 500 μm and PVDF concentration of 12% (*w/w*).

RH (%)	Exposure Time (min)	MPS (μm)	BP (μm)	CA (°)	LEP (kPa)
-	2	0.057	0.10	91.0	118.34
60%	5	0.060	0.13	94.3	167.66
-	10	0.137	0.78	104.5	137.18
-	2	0.073	0.17	93.6	141.17
80%	5	0.126	0.85	97.7	108.86
-	10	0.140	2.90	98.2	105.66

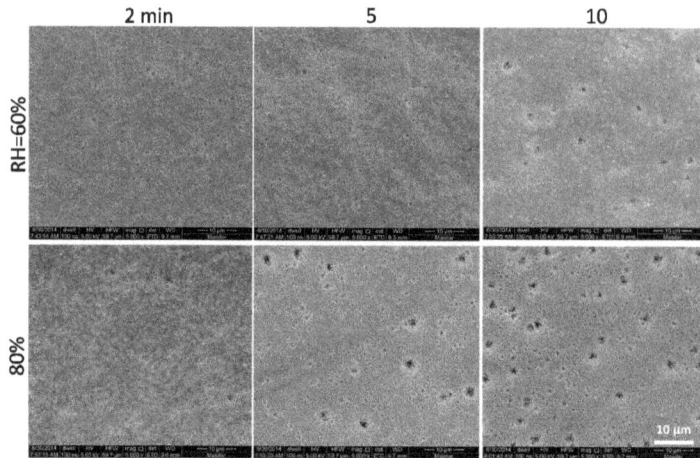

Figure 5. Surface SEM images of membrane samples prepared under different RH and different exposure times. Casting thickness: 500 μm, polymer solution: 12% *w/w* PVDF.

3.2.3. Effect of Exposure Time

A clear gradual change of the surface structure of membranes prepared using 12% PVDF, 60% RH and casting thickness of 500 μm is shown in Figure 6. This change translated into the in trend of CA shown in Figure 7A. A drastic change of surface structure was observed at exposure time between 1 and 2 h, in which the structure changed from a flat surface to a spherulitic-like structure (Figure 6). This structure is responsible for the substantial increase in CA up to 137°. However, the increase in CA was also mirrored by an increase in BP (Figure 7), which diminished the LEP value. The LEP data shows a gradual decrease with exposure time (Table 4), which is in agreement with the SEM and the BP results (Figures 6 and 7).

Increasing exposure time enlarged the membranes' MPS, which reached a plateau at 1 h, then jumped substantially at 2 h (Figure 7B). This is mainly because of the drastic change of membrane structure discussed above. Also, significant increase of BP only occurred after a certain exposure time (30 min). These findings suggest that extending the exposure time can significantly change the surface hydrophobicity without substantially changing the BP. In other words, it is possible to increase the LEP and MPS, which implies a larger MD flux at an assumed constant pore density, by extending the exposure time.

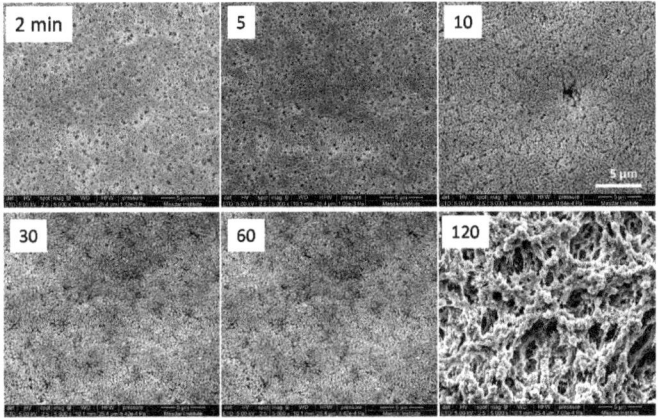

Figure 6. Surface SEM images of membrane samples prepared at different exposure times under 60% RH, 12% w/w PVDF concentration and 500-μm casting thickness. All images have the same scale bar, shown in the top-right image.

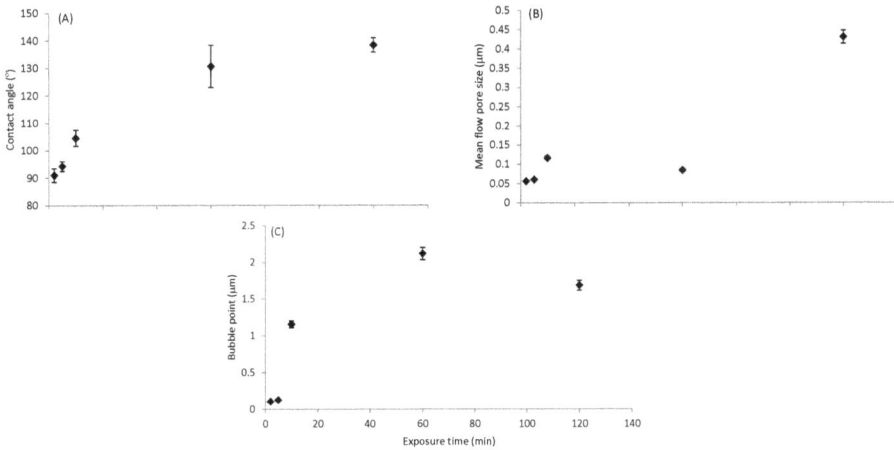

Figure 7. Impact of exposure time on (**A**) CA (contact angle); (**B**) MPS (mean flow pore size) and (**C**) BP (bubble-point pore size), for membranes prepared under RH of 60% and at casting thickness of 500 μm with 12% w/w PVDF concentration.

3.3. Moisture Uptake during the VIPS Process

Figure 8 shows mass variation of polymer cast film during exposure to different humidity values (37%, 60% and 80% RH). In all cases, soon after the exposure started, the film mass started to increase until it reached a maximum value, and then it decreased. As the RH increased, a shift in the maximum uptake was observed, peaking at around 20, 90 and 450 min for 37%, 60% and 80% RH, respectively. The rate of moisture uptake before reaching the maximum value was also faster for higher RH. Two competing mass transfer phenomena in the cast film are expected during the exposure: (1) the water flows into the film due to the hygroscopic and miscible nature of water and DMAC; and (2) the outflow of DMAC and water from the cast film. Before reaching the maximum mass, water uptake was a predominant phenomenon, leading to an increase in mass of the cast film. The increase in absorption rate of water at higher RH was a result of a larger chemical potential at higher water concentration in air. After reaching the maximum value, solvent and water evaporation from the casted film prevailed over water sorption, resulting in a weight loss. The water sorption stopped when there was no chemical potential difference, which occurred when the amount of water on the cast film was large enough. The loss of mass then continued as a result of the concomitant DMAC and water loss from the film.

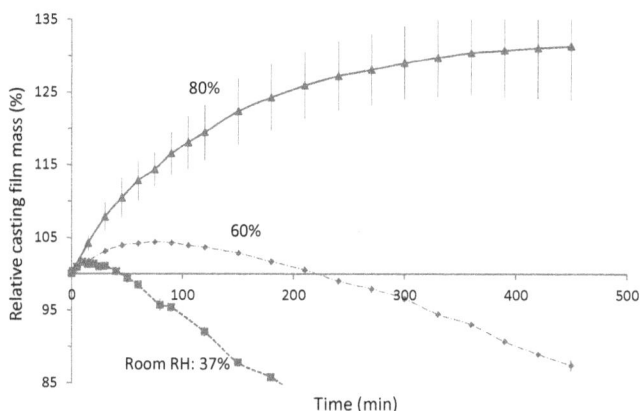

Figure 8. Water uptakes from humid atmosphere at different relative humidity (RH) values (37%, 60% and 80%) for membranes prepared using 12% *w/w* PVDF solution at a casting thickness of 500 μm.

The water sorption mechanism explained earlier allows the understanding of mass transfer phenomenon within the cast film. Just after the exposure started, at the very top of the film, a local exchange of water and DMAC occurred. This process continued further at deeper locations within the film creating a change in composition across the film thickness, as explained using a phenomenological model by [25]. This process would continue as long as the top film layer did not solidify as a result of phase separation. However, in the case of PVDF/DMAC/water system, during the humid air exposure, a thin film of water was observed on the top of the cast film after certain time, which was more obvious for higher RH (60% and 80%) as also found by [26] for a poly(ether-imide)/*N*-methylpyrolidone/water system. This finding suggests that the phase inversion of the top film layer had started and completed earlier, creating a dense barrier for further exchange of DMAC and water at the deeper location. Consequently, concentration gradient from top (being the lowest PVDF concentration) to the bottom (being the highest) could not be achieved fully. This explains why the increment of surface hydrophobicity only occurred within certain period of time (2–10 min) (Figure 7). Within this time interval and until just before phase separation completed, a rapid composition change within the cast film was still possible, resulting in a change in surface properties [10]. Up to this point, it is not clear why further exposure time beyond 1 h resulted in a shift of surface morphology as shown in Figure 6.

3.4. Decoupling Pore Size and Surface Properties: Effect of Casting Thickness

The three main parameters explored in Section 3.2 and the phenomenological explanation of water imbibition and water/DMAC mass transfer in Section 3.3 are important to (1) determine the properties of the resulting membrane and (2) the suitability of the membrane for MD application. In an attempt to elucidate the roles of those parameters and the underlying mechanisms to produce desired MD membranes, the impact of casting thickness was further explored. Our key hypothesis here is that the gradient in polymer concentration during the phase inversion process can be exploited by membrane thickness control, to obtain highly hydrophobic surface properties while maintaining an acceptable BP and MPS. The tests were performed under 80% RH, which offers higher chemical potential of water absorption into the cast film.

A gradual change in the surface morphology of the membrane samples was observed by extending the exposure time (Figure 9), which was similar to the one observed by varying the time parameter at RH of 60% in Figure 6. However, no obvious impact of casting thickness was observed, or at least it was not clearly visible from the surface SEM images. The change of CA with casting thickness shows a general decrease and this is clearly observed in the case of 7.5-min exposure time (Figure 10). It is also observed that the effect of casting thickness on CA and BP is more pronounced in the cases of relatively longer exposure times (5 and 7 min) in comparison to shorter ones (2 and 3.5 min). These observations may be attributed to the influx and outflux of water vapor and DMAC, respectively. At a thinner casting thickness, the resistance to mass transfer of water vapor and DMAC is relatively lower than that at a thicker casting thickness. Consequently, and due to this larger influx and outflux in the case of thinner casting thickness, the morphological change (and consequently roughness) is more pronounced, leading to higher contact angles. The consequences of lower mass transfer resistance in thinner membranes leading to larger influx and outflux of water vapor and DMAC, respectively, can explain the general decrease of BP with thickness. As more water vapor is absorbed by the cast film, more DMAC is partitioned out, leaving behind large void spaces and consequently pores. This effect is less prominent in thicker membranes, causing lower BP due to the higher mass transfer resistance to water vapor influx. In fact, these observations reconfirm the findings on the effect of increasing humidity on BP (Table 4). As was seen before, the BP increased with humidity due to the higher chemical potential between the cast solution and humid atmosphere. Therefore, by decreasing the mass transfer resistance or increasing the chemical potential (humidity), a larger BP is obtained. As a consequence of the CA and BP trends, the LEP values in Figure 10 show a slight increase in LEP with thickness and this is clearly seen for the case of the 5-min exposure time. In this case, the slight increase in LEP can potentially be attributed to the relatively larger percentage decrease in BP in comparison to the decrease in CA. As shown in Figure 10, the casting thickness played a significant role in maintaining a small BP. As the casting thickness was reduced, the incremental BP size significantly increased, which substantially diminished the LEP. Under this situation, increasing thickness offers substantial advantages since under thicker casting thickness (e.g., 1000 μm) LEP could be controlled well.

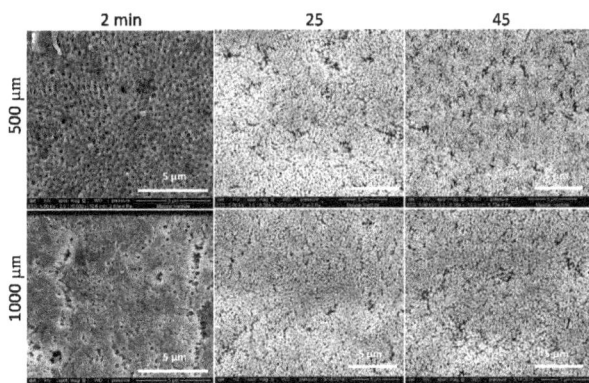

Figure 9. Surface SEM images of PVDF membranes prepared using VIPS process at different casting thicknesses and exposure times, all using 12% *w/w* PVDF polymer solution and under 80% RH.

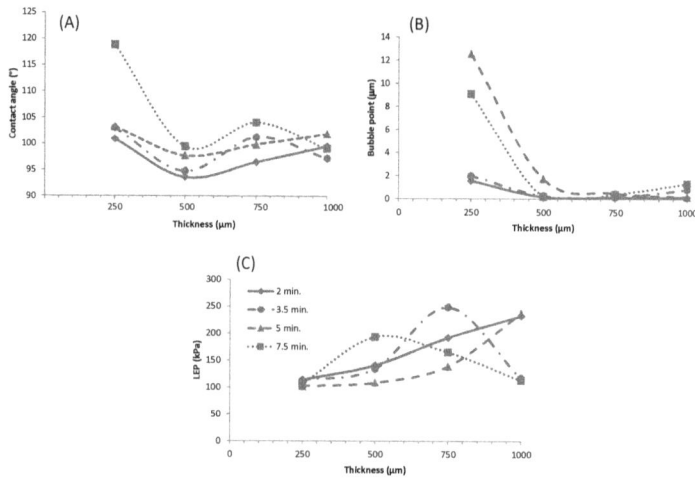

Figure 10. Impact of casting thickness on the properties of PVDF membrane prepared using VIPS process, all using 12% *w/w* PVDF polymer solution and under 80% RH. (**A**) Contact angle profile; (**B**) Bubble point profile; (**C**) LEP profile.

3.5. MD Performance

DCMD performance tests were performed for selected membranes with varying casting thicknesses (500 and 1000 μm) and exposure times (5, 10 and 15 min). These particular membranes were selected because they offered an acceptable range of contact angle and pore sizes (Table 5). The results (Table 5) showed a flux range of 9–32 L/m^2·h and SR > 99%. These flux values are within the range of MD flux reported for PVDF membranes [27–29]. In general, no obvious trend could be extracted from the salt rejection data since they are considerably high for all samples. When comparing the impact of casting thickness, at exposure time of 5 and 10 min, the flux of the 1000-μm membrane was higher than the 500-μm one, while the opposite trend was encountered at the 15-min exposure time. The trend of flux with exposure time for both casting thicknesses was also different. For the 500-μm thickness, the flux changed with exposure time: 8.75 (at 5 min), 13.77 (at 10 min) and 22.19 L/m^2·h (at 15 min). Meanwhile, for the 1000-μm membranes, the flux values were 27.76, increasing to 31.82 and then dropping to 10.98 for exposure times of 5, 10 and 15 min, respectively. The trend of flux values unfortunately cannot be very well explained with the existing data in this study. It is not a function of MPS, nor a function casting thickness. In addition to surface morphology and MPS, it is most likely that the VIPS process also alters other properties (e.g., tortuosity, pore density, surface porosity or other) that affect the mass transfer resistance of the resulting membranes. Nevertheless, the ability to obtain DCMD membranes having an acceptable rejection and flux while improving their LEP has been well proven in this study.

Table 5. Summary of direct-contact membrane distillation (DCMD) performance using membrane samples fabricated at 80% RH and 12% *w/w* PVDF casting solution.

Casting Thickness (μm)	Exposure Time (min)	Average Flux (L/m^2·h)	Average Salt Rejection (SR) (%)	Porosity (%)	MPS (μm)	Net Thickness (μm)
	5	8.7 ± 3.2	99.95	71	0.0863	246
500	10	13.8 ± 7.0	99.75	87	0.2347	265
	15	22.2 ± 6.9	99.91	76	0.1649	294
	5	27.8 ± 1.9	99.81	65	0.0693	307
1000	10	31.8 ± 2.9	99.91	68	0.0641	327
	15	11.0 ± 7.5	99.99	66	0.068	381

4. Conclusions

Through a study of VIPS and NIPS processes, it has been shown that the VIPS process is an effective method to obtain a PVDF membrane with high contact angle and low BP, through manipulation of casting thickness. The membrane properties were tuned in this study by exploiting the role of casting thickness to obtain a membrane with a high CA, low BP and acceptable DCMD flux and SR. The parametric study results show that decreasing the PVDF concentration of the polymer cast solution and increasing RH and exposure time increased CA, MPS and BP. Those trends were exploited further by varying the casting thickness. At higher casting thicknesses and longer exposure times, CA increased but BP significantly decreased, while the latter showed a more dominant trend leading to an optimum condition with respect to maximum LEP. This was obtained in this study at a higher casting thickness (1000 μm, thanks to lower BP) and short exposure time (5 min, thanks to higher CA). Moisture uptake experiments showed peaks at which the mass transfer mechanisms shifted from being dominated by water vapor absorption to evaporation of DMAC and water. This allowed the determination of the ideal exposure time to obtain the desired membrane hydrophobicity and pore size. The SR and DCMD fluxes of the selected membrane samples showed that they are within the value ranges of PVDF membranes reported for MD in literature, demonstrating a substantial gain in LEP without sacrificing flux, or rejection.

Acknowledgments: This work was funded by the Cooperative Agreement between the Masdar Institute of Science and Technology, Abu Dhabi, UAE and the Massachusetts Institute of Technology (MIT), Cambridge, MA, USA, Reference No. 02/MI/MI/CP/11/07633/GEN/G/00.

Author Contributions: F.A.A., M.R.B. and H.A.A. conceived and designed the experiments; F.A.A. and M.R.B. performed the experiments; F.A.A., M.R.B. and H.A.A. analyzed the data; H.A.A. contributed reagents/materials/analysis tools; F.A.A., M.R.B. and H.A.A. wrote the paper.

Conflicts of Interest: The authors declare no conflict of interest.

References

1. Warsinger, D.M.; Swaminathan, J.; Guillen-Burrieza, E.; Arafat, H.A.; Lienhard, J.H., V. Scaling and fouling in membrane distillation for desalination applications: A review. *Desalination* **2015**, *356*, 294–313. [CrossRef]
2. Mu, C.; Su, Y.; Sun, M.; Chen, W.; Jiang, Z. Fabrication of microporous membranes by a feasible freeze method. *J. Membr. Sci.* **2010**, *361*, 15–21. [CrossRef]
3. Peng, Y.; Fan, H.; Ge, J.; Wang, S.; Chen, P.; Jiang, Q. The effects of processing conditions on the surface morphology and hydrophobicity of polyvinylidene fluoride membranes prepared via vapor-induced phase separation. *Appl. Surf. Sci.* **2012**, *263*, 737–744. [CrossRef]
4. Simone, S.; Figoli, A.; Criscuoli, A.; Carnevale, M.C.; Rosselli, A.; Drioli, E. Preparation of hollow fibre membranes from PVDF/PVP blends and their application in VMD. *J. Membr. Sci.* **2010**, *364*, 219–232. [CrossRef]
5. Peng, Y.; Fan, H.; Dong, Y.; Song, Y.; Han, H. Effects of exposure time on variations in the structure and hydrophobicity of polyvinylidene fluoride membranes prepared via vapor-induced phase separation. *Appl. Surf. Sci.* **2012**, *258*, 7872–7881. [CrossRef]
6. Devi, S.; Ray, P.; Singh, K.; Singh, P.S. Preparation and characterization of highly micro-porous PVDF membranes for desalination of saline water through vacuum membrane distillation. *Desalination* **2014**, *346*, 9–18. [CrossRef]
7. Peng, M.; Li, H.; Wu, L.; Zheng, Q.; Chen, Y.; Gu, W. Porous poly(vinylidene fluoride) membrane with highly hydrophobic surface. *J. Appl. Polym. Sci.* **2005**, *98*, 1358–1363. [CrossRef]
8. Kuo, C.-Y.; Lin, H.-N.; Tsai, H.-A.; Wang, D.-M.; Lai, J.-Y. Fabrication of a high hydrophobic PVDF membrane via nonsolvent induced phase separation. *Desalination* **2008**, *233*, 40–47. [CrossRef]
9. Fan, H.; Peng, Y.; Li, Z.; Chen, P.; Jiang, Q.; Wang, S. Preparation and characterization of hydrophobic PVDF membranes by vapor-induced phase separation and application in vacuum membrane distillation. *J. Polym. Res.* **2013**, *20*, 134. [CrossRef]

10. Matsuyama, H.; Teramoto, M.; Nakatani, R.; Maki, T. Membrane formation via phase separation induced by penetration of nonsolvent from vapor phase. I. Phase diagram and mass transfer process. *J. Appl. Polym. Sci.* **1999**, *74*, 159–170. [CrossRef]

11. Felder, R.M.; Rousseau, R.W. *Elementry Principles of Chemical Processes, (with CD)*; John Wiley & Sons: Hoboken, NJ, USA, 2008.

12. Gopal, R.; Rizvi, S. Vapour pressures of some mono-and di-alkyl substituted aliphatic amides at different temperatures. *J. Indian Chem. Soc.* **1968**, *45*, 13–16.

13. Guillen-Burrieza, E.; Thomas, R.; Mansoor, B.; Johnson, D.; Hilal, N.; Arafat, H. Effect of dry-out on the fouling of PVDF and PTFE membranes under conditions simulating intermittent seawater membrane distillation (SWMD). *J. Membr.Sci.* **2013**, *438*, 126–139. [CrossRef]

14. Li, C.-L.; Wang, D.-M.; Deratani, A.; Quémener, D.; Bouyer, D.; Lai, J.-Y. Insight into the preparation of poly(vinylidene fluoride) membranes by vapor-induced phase separation. *J. Membr. Sci.* **2010**, *361*, 154–166. [CrossRef]

15. Dorrer, C.; Rühe, J. Some thoughts on superhydrophobic wetting. *Soft Matter* **2009**, *5*, 51–61. [CrossRef]

16. Feng, X.J.; Jiang, L. Design and Creation of Superwetting/Antiwetting Surfaces. *Adv. Mater.* **2006**, *18*, 3063–3078. [CrossRef]

17. Zuo, J.; Chung, T.-S. Metal–Organic Framework-Functionalized Alumina Membranes for Vacuum Membrane Distillation. *Water* **2016**, *8*, 586. [CrossRef]

18. Hilal, N.; Ismail, A.F.; Wright, C.J. PVDF Membranes for Membrane Distillation. In *Membrane Fabrication*; CRC Press: Boca Raton, FL, USA, 2015.

19. Khayet, M.; Matsuura, T. *Membrane Distillation: Principles and Applications*; Elsevier: Amsterdam, The Netherlands, 2011.

20. Strathmann, H.; Kock, K. The formation mechanism of phase inversion membranes. *Desalination* **1977**, *21*, 241–255. [CrossRef]

21. Li, M.; Katsouras, I.; Piliego, C.; Glasser, G.; Lieberwirth, I.; Blom, P.W.; de Leeuw, D.M. Controlling the microstructure of poly(vinylidene-fluoride)(PVDF) thin films for microelectronics. *J. Mater. Chem. C* **2013**, *1*, 7695–7702. [CrossRef]

22. Matsuyama, H.; Teramoto, M.; Nakatani, R.; Maki, T. Membrane formation via phase separation induced by penetration of nonsolvent from vapor phase. II. Membrane morphology. *J. Appl. Polym. Sci.* **1999**, *74*, 171–178. [CrossRef]

23. Wang, X.; Zhang, L.; Sun, D.; An, Q.; Chen, H. Formation mechanism and crystallization of poly(vinylidene fluoride) membrane via immersion precipitation method. *Desalination* **2009**, *236*, 170–178. [CrossRef]

24. Van de Witte, P.; Dijkstra, P.J.; Van den Berg, J.W.A.; Feijen, J. Phase separation processes in polymer solutions in relation to membrane formation. *J. Membr. Sci.* **1996**, *117*, 1–31. [CrossRef]

25. Caquineau, H.; Menut, P.; Deratani, A.; Dupuy, C. Influence of the relative humidity on film formation by vapor induced phase separation. *Polym. Eng. Sci.* **2003**, *43*, 798–808. [CrossRef]

26. Menut, P.; Su, Y.S.; Chinpa, W.; Pochat-Bohatier, C.; Deratani, A.; Wang, D.M.; Huguet, P.; Kuo, C.Y.; Lai, J.Y.; Dupuy, C. A top surface liquid layer during membrane formation using vapor-induced phase separation (VIPS)—Evidence and mechanism of formation. *J. Membr. Sci.* **2008**, *310*, 278–288. [CrossRef]

27. Alkhudhiri, A.; Darwish, N.; Hilal, N. Membrane distillation: A comprehensive review. *Desalination* **2012**, *287*, 2–18. [CrossRef]

28. Kang, G.; Cao, Y. Application and modification of poly(vinylidene fluoride) (PVDF) membranes—A review. *J. Membr. Sci.* **2014**, *463*, 145–165. [CrossRef]

29. Park, Y.-S.; Lee, C.-K.; Kim, S.-K.; Oh, H.-J.; Lee, S.-H.; Choi, J.-S. Effect of temperature difference on performance of membrane crystallization-based membrane distillation system. *Desalin. Water Treat.* **2013**, *51*, 1362–1365. [CrossRef]

applied
sciences

MDPI

Article

Increasing the Performance of Vacuum Membrane Distillation Using Micro-Structured Hydrophobic Aluminum Hollow Fiber Membranes

Chia-Chieh Ko [1], Chien-Hua Chen [1], Yi-Rui Chen [1], Yu-Hsun Wu [1], Soon-Chien Lu [1], Fa-Chun Hu [2], Chia-Ling Li [2] and Kuo-Lun Tung [1,*]

[1] Department of Chemical Engineering, National Taiwan University, Taipei 106, Taiwan;
    jackko0213@gmail.com (C.-C.K.); jefferychen1986@gmail.com (C.-H.C.);
    yi.rui.chen.rex@gmail.com (Y.-R.C.): b00504074@ntu.edu.tw (Y.-H.W.); durvlee2001@hotmail.com (S.-C.L.)
[2] Material and Chemical Research Laboratories, Industrial Technology Research Institute,
    Hsinchu 310, Taiwan; frankiehu@itri.org.tw (F.-C.H.); clli@itri.org.tw (C.-L.L.)
* Correspondence: kuolun@ntu.edu.tw; Tel.: +886-2-3366-3027; Fax: +886-2-2362-3040

Academic Editor: Enrico Drioli
Received: 29 January 2017; Accepted: 1 April 2017; Published: 4 April 2017

Abstract: This study develops a micro-structured hydrophobic alumina hollow fiber with a high permeate flux of $60 \, \text{Lm}^{-2}\text{h}^{-1}$ and salt rejection over 99.9% in a vacuum membrane distillation process. The fiber is fabricated by phase inversion and sintering, and then modified with fluoroalkylsilanes to render it hydrophobic. The influence of the sintering temperature and feeding temperature in membrane distillation (MD) on the characteristics of the fiber and MD performance are investigated. The vacuum membrane distillation uses 3.5 wt% NaCl aqueous solution at 70 °C at 0.03 bar. The permeate flux of $60 \, \text{Lm}^{-2}\text{h}^{-1}$ is the highest, compared with reported data and is higher than that for polymeric hollow fiber membranes.

Keywords: membrane distillation; vacuum membrane distillation; ceramic hollow fiber membrane; hydrophobicity; water desalination

1. Introduction

The availability of fresh water is limited. Even though almost 70% of the earth's surface is covered by water, more than 97% of water is in the oceans, leaving only 3% for consumption by human and animal life [1,2]. The rapid development of technology has shown that a supply of fresh water is critical. However, large amounts of energy are required to obtain fresh water using conventional equipment [3].

Membrane separation processes are an emerging technology that feature low energy consumption, especially membrane contactors with porous membranes [4,5]. Therefore, there have been many studies of membrane distillation (MD) processes in the last two decades [6–8]. MD is frequently used for seawater desalination due to the easy driving force of slight temperature differences [6,9–11]. Membrane contactors could resolve the fresh water problem. However, compared to conventional separation processes, there is extra resistance owing to the membrane in membrane contactors. In other words, the features of the membrane seriously affect the overall performance of membrane contactors.

At present, membranes used extensively in membrane contactors are organic polymer membranes, due to their higher porosity and thinness [12–14]. However, the poor chemical resistance and the ease with which organic polymer membranes swell make them unsuitable for chronic continuous operation. Few studies use inorganic ceramic membranes for membrane contactors because of lower porosity, which results from the sintering process that is used for ceramic membrane production. The hydrophilic nature of inorganic ceramic membranes also renders them inapplicable to membrane contactors. Inorganic ceramic membranes have potential in this area because of their outstanding thermal and chemical resistance and high mechanical strength [15–19]. The production of a hydrophobic, porous ceramic membrane is key to its success.

For membranes with various geometries, hollow fiber membranes have been widely studied because of their potential for low mass transfer resistance and high packing density [4,5,20]. Whether organic polymer or inorganic, ceramic hollow fiber membranes are generally fabricated using phase inversion, which is also a common way to produce the membrane for MD [3]. Phase inversion produces a hollow fiber with a symmetric or asymmetric structure that can be tuned by varying the operational parameters and the composition of the suspension, the bore fluid and coagulation [20–22]. This study fabricates hydrophobic ceramic fiber membranes for vacuum membrane distillation (VMD) using a high performance via phase-inversion/sintering method.

2. Experimental

2.1. Fabrication of Alumina Hollow Fiber Membranes

A combined phase-inversion and sintering method was used to fabricate the alumina hollow fiber membranes. The dispersant poly-ethyleneglycol 30-dipolyhydroxystearate (ARLACEL-P135, Croda Taiwan, Taoyuan, Taiwan, molecular weight: 5000 g/mol) was firstly dissolved in N-methyl-2-pyrrolidone (NMP) (TEDIA, Echo Chemical, Taipei, Taiwan, purity > 99%) and stirred to form a homogeneous solution. Aluminum oxide powder (Alfa Aesar, Uni-onward, New Taipei, Taiwan, α-phase > 99.9%, average particle size: 1.0 μm) and polyethersulfone (PESf) (Veradel A-301, SOLVAY, Trump Chemical, Taipei, Taiwan, amber color) were then added into the solution gradually and stirred for 48 h to form a continuous suspension. The weight percentages of aluminum oxide powder, polymer PES$_f$, solvent NMP and dispersant ARLACEL-P135 in the suspension were 49.5 wt%, 9.9 wt%, 39.6 wt%, and 0.99 wt%, respectively. The as-prepared suspension was transferred to a plastic injector and then extruded through a spinneret (outer diameter 2.0 mm, inner diameter 1.0 mm) from the injector, using a syringe pump. Deionized water was used as a bore fluid and pumped to the spinneret. After passing an air gap, the extruded suspension entered into a coagulation bath of deionized water that was used as a non-solvent for further solvent exchange, to obtain green fiber. This deionized water was renewed every 12 h to ensure a fresh non-solvent supply over a period of 2 days. The details of the spinning parameters are shown in Table 1.

After drying at room temperature, the as-prepared green hollow fiber was heated in a furnace at a rate of 1.6 °C/min to 480 °C and held at 480 °C for 12 h, followed by heating to specific temperature (1400 or 1500 °C) at a rate of 2 °C/min. It was then maintained for 2 h. The entire sintering process was performed in an air environment.

Table 1. The spinning parameters used in this study.

Parameters	Value
Suspension flow rate (mL/min)	15
Bore fluid flow rate (mL/min)	10
Air gap (cm)	20

2.2. Fiber Hydrophobization

Fluoroalkylsilanes (FAS) (1H, 1H, 2H, 2H-perfluorooctyltriethoxysilane) (Echo Chemical, Taipei, Taiwan, purity > 95%, molecular weight: 510.36 g/mol) as a hydrophobilizing agent were used to attach with the hydroxyl group on the as-sintered hollow fiber using an immersion method. The as-sintered hollow fiber was immersed into a 0.02 M FAS solution of n-hexane, which was the same concentration used in previous work, at 40 °C for 48 h and the FAS solution was refreshed every 24 h. This was followed by drying at room temperature, to produce a hydrophobic fiber. Finally, the hydrophobic fibers were stored in an ambient atmosphere followed by module procedure and MD.

2.3. Characterization

The membrane's hydrophobicity was quantified by measuring the apparent contact angle (CA) value before and after hollow fiber hydrophobization, to determine the hydrophobicity of the fiber surface.

A Scanning Electron Microscope (SEM) was utilized to observe the morphology and microstructure of the hollow fiber.

Hollow fibers were characterized by an extrusion method, using a mercury porosimeter to determine the porosity, the average pore size and the distribution of the pore size.

Pure water permeation was carried out using a home-made device before and after hollow fiber hydrophobization. One end of the hollow fiber was sealed with epoxy resin and then fixed at the outlet of a stainless pipe and covered. Deionized water was fed into the stainless pipe and pressurized by nitrogen at various pressures. The weight of the liquid deionized water penetrating from outside to inside through the hollow fiber was recorded using an electronic balance.

Liquid entry pressure of water (LEP_w), defined as the pressure at which water penetrates through the hydrophobic hollow fiber, used the same device as that used for pure water permeation. The LEP_w value is a critical parameter for the hydrophobicity level of the hollow fiber and determines whether it is applicable for the MD process.

2.4. Fiber Module

The module procedure of hydrophobic fiber was required before membrane distillation. The fiber module contained a hydrophobic hollow fiber and glass shell. The hollow fiber with a length of 9 cm was incorporated into the glass shell with a 1.8 cm inner diameter and a length of 10.4 cm using epoxy resin. There was one inlet and one outlet for the shell side and only one outlet for the permeate side, whereas the other side of fiber was sealed, as shown in Figure 1.

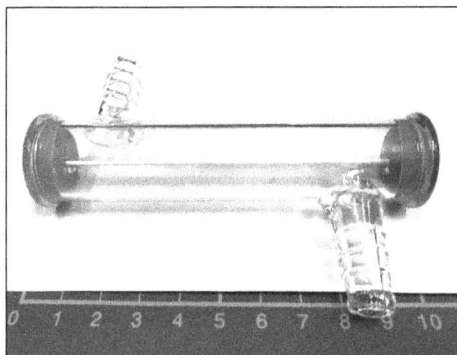

Figure 1. Hydrophobic aluminum hollow fiber module.

2.5. Membrane Distillation

Membrane performance was assessed using a home-made VMD configuration with a hollow fiber module. NaCl aqueous solution at a concentration of 3.5 wt% was heated to 70 °C and circulated at a flow rate of 1 L/min into the shell side, as the feed side of the module. The pressure of the lumen side, which is generally called the permeate side, was maintained at 0.03 bar using the vacuum pump. The water vapor that was transported from the shell side to the lumen side of the module was condensed and collected in a trap bottle using liquid nitrogen. The permeate flux and the rejection of the VMD process were determined by measuring the weight as a function of time and the conductivity of the condensing water, respectively.

3. Results and Discussion

The general temperature used in the sintering process to fabricate the ceramic hollow fiber membrane is about 1300 to 1500 °C, therefore the sintering temperatures of 1400 and 1500 °C are investigated in this study. SEM analysis confirmed the morphologies of the as-sintered alumina hollow fiber membrane sintered at 1400 and 1500 °C. As shown in Figure 2a,b, the alumina grain size of the outer surface sintered at 1400 °C is slightly smaller than that at 1500 °C, and for both the outer surface consists of visible alumina grains and a preliminary result for the pore size of several hundred micrometer are evident. From the cross-sectional SEM images (Figure 2c,d) of the as-sintered alumina hollow fiber membrane, it is seen that a symmetrical fiber with an inner and outer diameter of 0.8 and 1.2 mm for 1400 °C sintering and that of 0.8 and 1.3 mm for 1500 °C sintering are produced. Figure 2e,f both show the typical microstructure that results from the phase-inversion method. For both fabrications sintered at 1400 and 1500 °C, a finger-like structure extends from two surfaces and a sponge like structure fills the middle part, which shows that the spinning procedure is successful. The finger-like structure, which occupies at least 70% of the entire structure, observed in Figure 2e, is slightly more than that observed in Figure 2f, and the more finger-like structure may result in a better permeating characteristic.

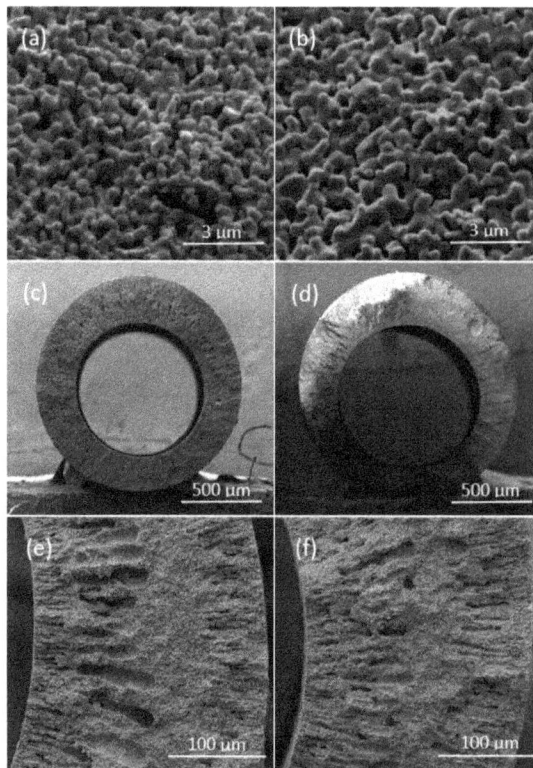

Figure 2. The morphologies of the as-sintered alumina hollow membrane sintered at different temperatures: outer surface sintered at (**a**) 1400 °C and (**b**) 1500 °C; cross-sectional sintered at (**c**) 1400 °C and (**d**) 1500 °C; partial cross-sectional sintered at (**e**) 1400 °C and (**f**) 1500 °C.

The mercury porosimeter results for the as-prepared fiber are shown in Figure 3. It is seen that both of the pore size measurements are in the range of 100 to 400 nm, that give average pore size of about 220 and 165 nm for 1400 and 1500 °C sintering, respectively. They both are consistent with the appropriate pore size of 0.1 to 1 μm that has been reported for MD. In addition, the porosities of 55% and 33% were measured for the materials sintered at 1400 and 1500 °C, respectively.

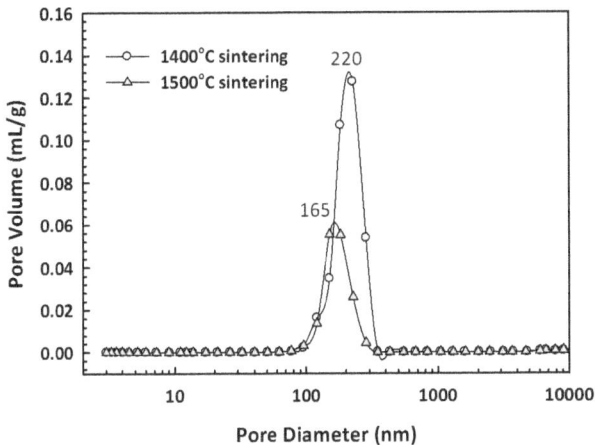

Figure 3. Pore size distribution, as measured by a mercury porosimeter.

The as-sintered hollow fiber is intrinsically hydrophilic because of the presence of hydroxyl groups on the fiber. In order to use the fiber for MD, FAS, which is a general agent for hydrophobization, was used to hydrophobilize the fiber by the method mentioned previously. A pure water permeation test was carried out before and after hydrophobization, to determine the permeability and the LEP_w value which determines the applicability for MD.

The results of the pure water permeation as a function of the pressure difference between the feed and permeate sides are shown in Figure 4. The pure water fluxes of the fiber sintered at 1400 and 1500 °C both increase almost linearly with pressure difference before hydrophobization, and pure water fluxes of about 3000 and 2600 $Lm^{-2}h^{-1}bar^{-1}$ are recorded. The higher pure water flux of 3000 $Lm^{-2}h^{-1}bar^{-1}$ is attributed to the larger average pore size (220 nm) and the higher porosity (55%) mentioned above. However, the water flux could not be detected after hydrophobization until a specific pressure difference, which means the LEP_w value. The LEP_w value of 2.5 bar for 1400 °C sintering (Figure 4c) is lower than that of 4.5 bar for 1500 °C sintering (Figure 4d), and this result is in accordance with the larger average pore size of 220 nm. Both LEP_w values are higher than the pressure difference of about 1.0 bar used in VMD, which also demonstrates the applicability. Figure 4e,f also shows the hydrophobicity with contact angles both at about 137°. Furthermore, it has been reported that the FAS modification only reduces the pore size of the fiber slightly for several nanometers.

After characterization, the as-prepared hydrophobic fiber modules were used for MD. VMD was used because of its high permeate flux. The 3.5 wt% NaCl aqueous solution was circulated at the shell side and the permeate water was collected at the tube side at 0.03 bar every 30 min. As shown in Figure 5, the permeate flux of the fiber sintered at 1400 °C is twice as high compared to that sintered at 1500 °C, and this result is owing to the higher average pore size and porosity of the former. The salt rejections, which were detected by measuring conductivity, are both greater than 99.9%. In addition, the feeding temperature dependence of the permeate flux is investigated with the as-prepared fiber sintered at 1400 °C. As shown in Figure 6, the permeate flux increases with the temperature of feed solution in the range of 50 °C to 70 °C, that is expected from the relation between temperature and vapor pressure which is the driving force in VMD. The high permeate flux of 60 $Lm^{-2}h^{-1}$ is obtained at a feed temperature of 70 °C and the salt rejections under three of the feed temperatures are all greater than 99.9% over a period of 4.5 h. These results demonstrate that the as-prepared hydrophobic fiber is eminently suitable for VMD. A list of the permeate flux values for ceramic membranes published recently is summarized in Table 2. The permeate flux depends principally on the feed solution and the configuration. A variety of configurations including direct contact membrane distillation (DCMD), air-gap membrane distillation (AGMD), sweeping-gas membrane distillation (SGMD) and vacuum membrane distillation (VMD) have been used for MD. They can be distinguished by the driving method inclusive of a condensing fluid contacting with the membrane (DCMD), an air gap between the condensing surface and membrane (AGMD), a sweeping gas (SWMD) and a vacuum condition at the permeate side. The choice of configuration used for MD is dependent on the feeding composition, volatility, flux and operation. DCMD is easy to operate and needs the least equipment. VMD generally exhibits a high permeate flux due to the driving from vacuum. SGMD is usually used to remove a volatile organic from an aqueous solution. AGMD is widely used due to its characteristic of collecting separately at permeate side. All of them can be applied to seawater desalination. For the purpose of comparison, the permeate flux is plotted as a function of feed solution temperature in Figure 7. It is seen that the permeate flux value is much greater than those in the published data using a ceramic membrane with various configurations. The greater permeate flux shown in this study is attributed to the appropriate pore size (Figure 3), high porosity and the symmetrical finger-like structure (Figure 4e), which provides macro voids that enhance the mass transfer. On the other hand, it is believed that high contact angle, well permeation and the appropriate LEP_w value are indispensable to MD. Furthermore, checking the durability of the as-prepared hollow fiber over longer periods will be the subject of future work.

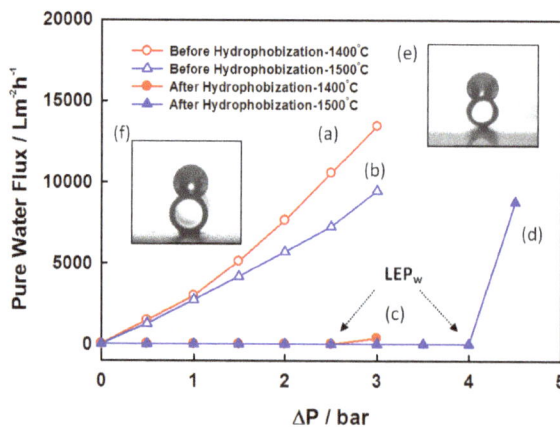

Figure 4. Pure water permeation for as-prepared alumina hollow membrane: before hydrophobization/sintered at (**a**) 1400 °C, (**b**) 1500 °C; after hydrophobization /sintered at (**c**) 1400 °C, (**d**) 1500 °C; contact angle after hydrophobization/sintered at (**e**) 1400 °C, (**f**) 1500 °C.

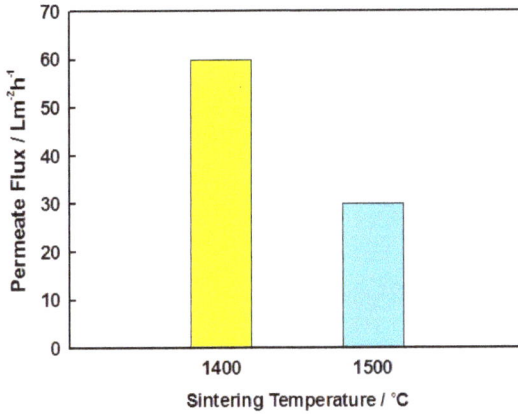

Figure 5. The membrane distillation (MD) permeate flux of as-prepared hydrophobic hollow fiber sintered at different temperatures in vacuum membrane distillation (VMD).

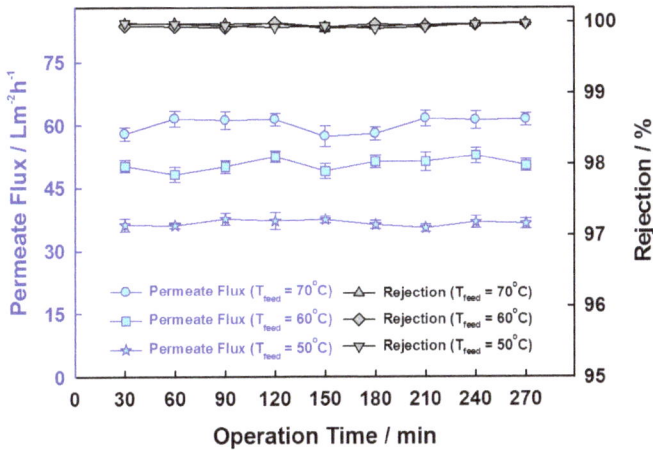

Figure 6. The performance of as-prepared hydrophobic hollow fiber sintered at 1400 °C with different feeding temperature (70, 60 and 50 °C) in VMD.

Table 2. A list of the MD permeate flux values for ceramic membranes published recently.

Configuration	Membrane Material	Membrane Morphology	NaCl Feed Solution	Hot Side Temp (°C)	MD Permeate Flux ($Lm^{-2}h^{-1}$)	Ref.
SGMD	Al_2O_3	Disk	4.0 wt%	70 °C (80 °C) [90 °C]	9.9 (13.1) [19.8]	[8]
AGMD	ZrO_2/Al_2O_3	Tubular	1.0 [M]	65 °C (80 °C) [95 °C]	1.0 (2.9) [6.9]	[23]
AGMD	Al_2O_3/ZrO_2	Tubular	1.0 [M]	75 °C (85 °C) [95 °C]	1.7 (2.9) [5.0]	[24]
AGMD	ZrO_2	Tubular	0.9 wt%	50 °C (60 °C) [70 °C]	2.3 (3.9) [7.0]	[25]
AGMD	Titania	Tubular	0.8 [M]	70 °C (80 °C) [90 °C]	0.8 (1.5) [2.8]	[26]
DCMD	Titania	Tubular	0.8 [M]	70 °C (80 °C) [90 °C]	0.6 (1.0) [2.5]	[26]
DCMD	Al_2O_3	Planar	0.5 [M]	53 °C	9.0	[27]
VMD	ZrO_2/Ti	Tubular	0.1 [M]	40 °C	10.8	[27]
AGMD	ZrO_2/Ti	Tubular	0.5 [M]	75 °C (85 °C) [95 °C]	2.7 (3.3) [4.7]	[27]
DCMD	Al_2O_3	Planar	4.0 wt%	60 °C (70 °C) [80 °C]	7.1 (11.5) [17.0]	[28]
DCMD	Al_2O_3/ZrO_2	Tubular	1.0 [M]	60 °C (95 °C)	1.1 (6.9)	[29]

Table 2. *Cont.*

Configuration	Membrane Material	Membrane Morphology	NaCl Feed Solution	Hot Side Temp (°C)	MD Permeate Flux (Lm^{-2}h^{-1})	Ref.
VMD	Si$_3$N$_4$	Hollow fiber	4.0 wt%	60 °C (70 °C) [80 °C]	14.6 (22.9) [27.5]	[30]
DCMD	Si$_3$N$_4$	Hollow fiber	4.0 wt%	60 °C (70 °C) [80 °C]	5.4 (7.5) [10.4]	[30]
VMD	Al$_2$O$_3$	Planar	4.0 wt%	60 °C (70 °C) [80 °C]	5.8 (7.9) [10.4]	[31]
DCMD	β-Sialon	Hollow fiber	4.0 wt%	60 °C (70 °C) [80 °C]	3.8 (5.4) [6.7]	[31]
VMD	Al$_2$O$_3$	Hollow fiber	4.0 wt%	60 °C (70 °C) [80 °C]	19.0 (32.0) [42.9]	[32]
VMD	Zeolite/Al$_2$O$_3$	Tubular	3.5 wt%	60 °C	12.0	[33]
DCMD	ZrO$_2$/Ti	Tubular	0.5 [M]	75 °C (85 °C) [95 °C]	1.7 (2.5) [3.8]	[34]
VMD	Si$_3$N$_4$	Hollow fiber	4.0 wt%	50 °C (60 °C) [70 °C]	9.6 (14.6) [22.2]	[35]
SGMD	Si$_3$N$_4$	Planar	4.0 wt%	75 °C	6.7	[36]
VMD	Al$_2$O$_3$	Disk	3.5 wt%	70 °C	37.1	[37]
DCMD	TiO$_2$	Nanofiber	3.5 wt%	60 °C (70 °C) [80 °C]	7.1 (9.2) [11.9]	[38]
VMD	Al$_2$O$_3$	Hollow fiber	3.5 wt%	50 °C (60 °C) [70 °C]	35.0 (47.0) [60.0]	This work

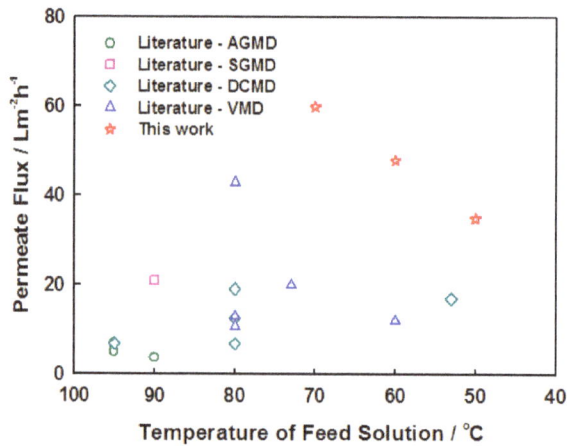

Figure 7. A comparison between the MD permeate flux value obtained from this work and historical data, as a function of feed solution temperature.

4. Conclusions

A hydrophobic alumina hollow fiber with high performance in VMD is successfully prepared. The alumina hollow fiber membrane is fabricated using a method that combines phase-inversion and sintering. The as-sintered hollow fiber is hydrophobilized using fluoroalkylsilanes, to obtain a hydrophobic hollow fiber. The hydrophobic hollow fiber prepared at 1400 °C has a thickness of 230 μm, an average pore size of 220 nm, porosity of 55%, a contact angle of 137° and an LEP$_w$ value of 2.5 bar. It exhibits a high permeate flux value of 60 Lm^{-2}h^{-1} which is twice as high compared to that prepared at 1500 °C, and a salt rejection of greater than 99.9% in VMD that uses 3.5 wt% NaCl aqueous solution at 70 °C at 0.03 bar. The permeate flux value of 60 Lm^{-2}h^{-1} is much greater than previously reported values using ceramic membranes. This hydrophobic alumina hollow fiber is suitable for use in seawater desalination. Checking the durability of the as-prepared hollow fiber over a longer duration will be the subject of future work.

Acknowledgments: The authors would like to thank the Ministry of Science and Technology (MOST), Taiwan, for its financial support (Project numbers 104-2221-E-002-176-MY3, 105-2622-E-002-010-CC1 and 105-2923-E-002 -011-MY2) and acknowledge their gratitude for the collaboration project funding from the Industrial Technology Research Institute, Taiwan.

Author Contributions: C.-C.K., C.-H.C., F.-C.H., C.-L.L. and K.-L.T. conceived and designed the experiments; C.-C.K. performed the experiments; C.-C.K. and Y.-R.C. analyzed the data; Y.-R.C., Y.-H.W. and S.-C.L. contributed reagents/materials/analysis tools; C.-C.K. wrote the paper.

Appl. Sci. **2017**, *7*, 357

Conflicts of Interest: The authors declare no conflict of interest.

References

1. Shirazi, M.M.; Kargari, A.; Shirazi, M.J. Direct contact membrane distillation for seawater desalination. *Desalin. Water Treat.* **2012**, *49*, 368–372. [CrossRef]
2. Shiklomanov, I. World fresh water resources. In *Water in Crisis: A Guide to the World's Fresh Water Resources*; Gleick, P.H., Ed.; Oxford University Press, Inc.: New York, NY, USA, 1993.
3. Drioli, E.; Ali, A.; Macedonio, F. Membrane distillation: Recent developments and perspectives. *Desalination* **2015**, *356*, 56–84. [CrossRef]
4. Drioli, E.; Criscuoli, A.; Curcio, E. *Membrane Contactors: Fundamentals, Applications and Potentialities*; Elsevier Science: Amsterdam, The Netherlands, 2006.
5. Noble, R.D.; Stern, S.A. *Membrane Separations Technology: Principle and Applications*; Elsevier Science: Amsterdam, The Netherlands, 1995.
6. Alkhudhiri, A.; Darwish, N.; Hilal, N. Membrane distillation: A comprehensive review. *Desalination* **2012**, *287*, 2–18. [CrossRef]
7. Ma, Z.; Hong, L.; Su, M. Superhydrophobic membranes with ordered arrays of nanospiked microchannels for water desalination. *Langmuir* **2009**, *25*, 5446–5450. [CrossRef] [PubMed]
8. Gu, J.Q.; Ren, C.L.; Zong, X.; Chen, C.S.; Winnubst, L. Preparation of alumina membranes comprising a thin separation layer and a support with straight open pores for water desalination. *Ceram. Int.* **2016**, *42*, 12427–12434. [CrossRef]
9. Manawi, Y.M.; Khraisheh, M.A.M.M.; Fard, A.K.; Benyahia, F.; Adham, S. A predictive model for the assessment of the temperature polarization effect in direct contact membrane distillation desalination of high salinity feed. *Desalination* **2014**, *341*, 38–49. [CrossRef]
10. Lawson, K.W.; Loyd, D.R. Membrane distillation. *J. Membr. Sci.* **1997**, *124*, 1–25. [CrossRef]
11. Khayet, M. Membranes and theoretical modeling of membrane distillation: A review. *Adv. Colloid Interface Sci.* **2011**, *164*, 56–88. [CrossRef] [PubMed]
12. Srisurichan, S.; Jiraratananon, R.; Fane, A.G. Humic acid fouling in the membrane distillation process. *Desalination* **2005**, *174*, 63–72. [CrossRef]
13. Ding, Z.; Ma, R.; Fane, A.G. A new model for mass transfer in direct contact membrane distillation. *Desalination* **2003**, *155*, 205. [CrossRef]
14. Boubakri, A.; Hafiane, A.; Bouguecha, S.A.T. Nitrate removal from aqueous solution by direct contact membrane distillation using two different commercial membranes. *Desalin. Water Treat.* **2015**, *26*, 2723–2730. [CrossRef]
15. Sommer, S.; Melin, T. Performance evaluation of microporous inorganic membranes in the dehydration of industrial solvents. *Chem. Eng. Process.* **2005**, *44*, 1138–1156. [CrossRef]
16. Lin, Y.F.; Chen, C.H.; Tung, K.L.; Wei, T.Y.; Lu, S.Y.; Chang, K.S. Mesoporous fluorocarbon-modified silica aerogel membranes enabling long-term continuous CO_2 capture with large absorption flux enhancements. *ChemSusChem* **2013**, *6*, 437–442. [CrossRef] [PubMed]
17. Lin, Y.F.; Wang, C.S.; Ko, C.C.; Chen, C.H.; Chang, K.S.; Tung, K.L.; Lee, K.R. Polyvinylidene fluoride/siloxane nanofibrous membranes for long-term continuous CO_2-capture with large absorption-flux enhancement. *ChemSusChem* **2014**, *7*, 604–609. [CrossRef] [PubMed]
18. Lin, Y.F.; Ko, C.C.; Chen, C.H.; Tung, K.L.; Chang, K.S. Reusable methyltrimethoxysilane-based mesoporous water-repellent silica aerogel membranes for CO_2 capture. *RSC Adv.* **2014**, *4*, 1456–1459. [CrossRef]
19. Lin, Y.F.; Ko, C.C.; Chen, C.H.; Tung, K.L.; Chang, K.S.; Chung, T.W. Sol-gel preparation of polymethylsilsesquioxane aerogel membranes for CO_2 absorption fluxes in membrane contactors. *Appl. Energy* **2014**, *128*, 25–31. [CrossRef]
20. Li, K. *Ceramic Membranes for Separation and Reaction*; Wiley: Hoboken, NJ, USA, 2007.
21. Lee, M.; Wu, Z.; Wang, R.; Li, K. Micro-structured alumina hollow fiber membranes— Potential applications in wastewater treatment. *J. Membr. Sci.* **2014**, *461*, 39–48. [CrossRef]
22. Tan, X.; Liu, N.; Meng, B.; Liu, S. Morphology control of the perovskite hollow fiber membranes for oxygen separation using different bore fluids. *J. Membr. Sci.* **2011**, *378*, 308–318. [CrossRef]

23. Krajewski, S.R.; Kujawski, W.; Bukowska, M.; Picard, C.; Larbot, A. Application of fluoroalkylsilanes (FAS) grafted ceramic membranes in membrane distillation process of NaCl solutions. *J. Membr. Sci.* **2006**, *281*, 253–259. [CrossRef]

24. Gazagnes, L.; Cerneaux, S.; Persin, M.; Prouzet, E.; Larbot, A. Desalination of sodium chloride solutions and seawater with hydrophobic ceramic membranes. *Desalination* **2007**, *217*, 260–266. [CrossRef]

25. Khemakhem, M.; Khemakhem, S.; Amar, R.B. Emulsion separation using hydrophobic grafted ceramic membranes by. *Colloids Surf. A Physicochem. Eng. Asp.* **2013**, *436*, 402–407. [CrossRef]

26. Kujawa, J.; Cerneaux, S.; Koter, S.; Kujawski, W. Highly efficient hydrophobic titania ceramic membranes for water desalination. *Appl. Mater. Interfaces* **2014**, *6*, 14223–14230. [CrossRef] [PubMed]

27. Hendren, Z.D.; Brant, J.; Wiesner, M.R. Surface modification of nanostructured ceramic membranes for direct contact membrane distillation. *J. Membr. Sci.* **2009**, *31*, 1–10. [CrossRef]

28. Ren, C.L.; Fang, H.; Gu, J.Q.; Winnubst, L.; Chen, C.S. Preparation and characterization of hydrophobic alumina planar membranes for water desalination. *J. Eur. Ceram. Soc.* **2015**, *35*, 723–730. [CrossRef]

29. Larbot, A.; Gazagnes, L.; Krajewski, S.; Bukowska, M.; Kujawski, W. Water desalination using ceramic membrane distillation. *Desalination* **2004**, *168*, 367–372. [CrossRef]

30. Zhang, J.W.; Fang, H.; Wang, J.W.; Hao, L.Y.; Xu, X.; Chen, C.S. Preparation and characterization of silicon nitride hollow fiber membranes for seawater desalination. *J. Membr. Sci.* **2014**, *450*, 197–206. [CrossRef]

31. Wang, J.W.; Li, L.; Zhang, J.W.; Xu, X.; Chen, C.S. β-Sialon ceramic hollow fiber membranes with high strength and low thermal conductivity for membrane distillation. *J. Eur. Ceram. Soc.* **2016**, *36*, 59–65. [CrossRef]

32. Fang, H.; Gao, J.F.; Wang, H.T.; Chen, C.S. Hydrophobic porous alumina hollow fiber for water desalination via membrane distillation process. *J. Membr. Sci.* **2012**, *403–404*, 41–46. [CrossRef]

33. Garofalo, A.; Donato, L.; Drioli, E.; Criscuoli, A.; Carnevale, M.C.; Alharbi, O.; Aljlil, S.A.; Algieri, C. Supported MFI zeolite membranes by cross flow filtration for water treatment. *Sep. Purif. Technol.* **2014**, *137*, 28–35. [CrossRef]

34. Cerneaux, S.; Strużyńska, I.; Kujawski, W.M.; Persin, M.; Larbot, A. Comparison of various membrane distillation methods for desalination using hydrophobic ceramic membranes. *J. Membr. Sci.* **2009**, *337*, 55–60. [CrossRef]

35. Zhang, J.-W.; Fang, H.; Hao, L.-Y.; Xu, X.; Chen, C.-S. Preparation of silicon nitride hollow fibre membrane for desalination. *Mater. Lett.* **2012**, *68*, 457–459. [CrossRef]

36. Wang, J.-W.; Li, L.; Gu, J.-Q.; Yang, M.-Y.; Xu, X.; Chen, C.-S.; Wang, H.-T.; Agathopoulos, S. Highly stable hydrophobic SiNCO nanoparticle-modified silicon nitride membrane for zero-discharge water desalination. *AIChE J.* **2017**, *63*, 1272–1277. [CrossRef]

37. Huang, C.-Y.; Ko, C.-C.; Chen, L.-H.; Huang, C.-T.; Tung, K.-L.; Liao, Y.-C. A simple coating method to prepare superhydrophobic layers on ceramic alumina for vacuum membrane distillation. *Sep. Purif. Technol.* **2016**. [CrossRef]

38. Fan, Y.; Chen, S.; Zhao, H.; Liu, Y. Distillation membrane constructed by TiO$_2$ nanofiber followed by fluorination for excellent water desalination performance. *Desalination* **2017**, *405*, 51–58. [CrossRef]

applied sciences

MDPI

Article

Wetting Resistance of Commercial Membrane Distillation Membranes in Waste Streams Containing Surfactants and Oil

Lies Eykens [1,2,*], Kristien De Sitter [1], Chris Dotremont [1], Wim De Schepper [1], Luc Pinoy [3] and Bart Van Der Bruggen [2,4]

1   VITO—Flemish Institute for Technological Research, Boeretang 200, 2400 Mol, Belgium; kristien.desitter@vito.be (K.D.S.); chris.dotremont@vito.be (C.D.); wim.deschepper@vito.be (W.D.S.)
2   Department of Chemical Engineering, Katholieke Universiteit Leuven, Celestijnenlaan 200F, B-3001 Leuven, Belgium; bart.vanderbruggen@kuleuven.be
3   Department of Chemical Engineering, Cluster Sustainable Chemical Process Technology, KU Leuven, Gebroeders Desmetstraat 1, B-9000 Ghent, Belgium; luc.pinoy@kuleuven.be
4   Faculty of Engineering and the Built Environment, Tshwane University of Technology, Private Bag X680, Pretoria 0001, South Africa
*   Correspondence: lies.eykens@vito.be; Tel.: +32-14-33-5663

Academic Editor: Enrico Drioli
Received: 13 December 2016; Accepted: 20 January 2017; Published: 25 January 2017

Abstract: Water management is becoming increasingly challenging and several technologies, including membrane distillation (MD) are emerging. This technology is less affected by salinity compared to reverse osmosis and is able to treat brines up to saturation. The focus of MD research recently shifted from seawater desalination to industrial applications out of the scope of reverse osmosis. In many of these applications, surfactants or oil traces are present in the feed stream, lowering the surface tension and increasing the risk for membrane wetting. In this study, the technological boundaries of MD in the presence of surfactants are investigated using surface tension, contact angle and liquid entry pressure measurements together with lab-scale MD experiments to predict the wetting resistance of different membranes. Synthetic NaCl solutions mixed with sodium dodecyl sulfate (SDS) were used as feed solution. The limiting surfactant concentration was found to be dependent on the surface chemistry of the membrane, and increased with increasing hydrophobicity and oleophobicity. Additionally, a hexadecane/SDS emulsion was prepared with a composition simulating produced water, a waste stream in the oil and gas sector. When hexadecane is present in the emulsion, oleophobic membranes are able to resist wetting, whereas polytetrafluoretheen (PTFE) is gradually wetted by the feed liquid.

Keywords: membrane distillation; wetting; sodium dodecyl sulfate; hexadecane

1. Introduction

Membrane distillation (MD) is a thermally-driven membrane separation process, mostly applied to separate salts from an aqueous solution. The process uses a hydrophobic membrane to retain the liquid phase, while vapor is transported through the microporous structure. In direct contact membrane distillation, the membrane can be considered a contactor between the process liquids and the vapor phase, enabling 100% retention of dissolved components. Critical for the process is that the membrane pores are not wetted by the process liquids. As with any thermal separation process, MD is considered less energy efficient than reverse osmosis [1]. However, unlike reverse osmosis, MD can be applied using low-grade waste heat, solar or geothermal energy, considerably reducing the energy costs [2]. Moreover, the process flux and salt retention are less affected by salinity of the feed and therefore the process is able to operate properly up to saturation [3,4]. Recently, membrane distillation has been applied more and more for challenging water streams with a much higher load of contaminants compared to seawater desalination. Examples include reverse osmosis (RO) brines [5–7], industrial waste water [8,9] and produced water [10,11]. In many of the applications, the contaminants include a combination of less soluble salts, organic foulants or components lowering the surface tension of the process fluids. The occurrence of scaling and fouling of these contaminants can decrease the performance of the membrane [6,12–19]. Another less investigated phenomenon is membrane wetting, where despite the hydrophobicity and controlled pore size, liquid is able to penetrate into the membrane. The liquid entry pressure (LEP) indicates the minimum pressure difference over the membrane at which membrane wetting will occur and is given by [20]:

$$LEP = -\frac{2\gamma_L \cos\theta}{r_{max}} \tag{1}$$

where γ_L is the surface tension of the liquid, θ the contact angle of the liquid with the surface and r_{max} the maximum pore radius.

Multiple origins of membrane wetting are known. Different MD studies show that the reduced hydrophobicity (θ) due to membrane fouling is a possible cause of membrane wetting [21,22]. Furthermore, defects present in the membrane influence r_{max} and therefore strongly affect the liquid entry pressure [23,24]. The main focus of this article is the presence of organics lowering the surface tension (γ_L) of the feed stream, which can cause wetting as well [20,25,26]. The effect of membrane wetting for a series of alcohols, organic acids and solvents was already investigated [20,25]. Different approaches were used to quantify the wetting resistance of a membrane towards the presence of an organic component:

The determination of the concentration and surface tension (γ_{pd}) at which a droplet wets the membrane (penetrating drop method) for each organic component.

Determination of the theoretical maximum allowable surface tension in the process (γ_{pc}) using [25]:

$$\gamma_{pc} = \gamma_{pd} + \frac{\Delta P \cdot r_{max}}{2B} \tag{2}$$

Where ΔP is the pressure drop over the membrane, r_{max} the maximum membrane pore size and B a geometric factor between 0 and 1, where 1 indicates a perfectly cylindrical pore.

Measurement of the liquid entry pressure occurs as a function of the liquid composition.

Whereas the effect of alcohols, organic acids and solvent on the wetting behavior is described in literature [20,25], the effect of surfactants and oil traces on the membrane distillation performance has been less studied. Lin et al. [26] observed membrane wetting of a 0.45 µm polytetrafluoretheen (PTFE) membrane at a sodium dodecyl sulfate (SDS) concentration of 28 mg/L, while a chemically-modified omniphobic membrane was able to resist wetting up to 115 mg/L SDS. Wang et al. showed that a 1000 ppm crude oil mixture immediately wets a 0.45 µm polyvinylidene fluoride (PVDF) membrane [27], while 100 ppm mineral oil is shown to substantially wet the membrane after 21 h [28].

This study aims to improve the understanding of membrane wetting in the presence of surfactants and oil. A commonly used surfactant, sodium dodecyl sulfate, was used to investigate the effect on surface tension, contact angle and membrane wetting in membrane distillation. The validity of a simple methodology correlating the surface tension, contact angle and liquid entry pressure to membrane wetting in membrane distillation was investigated for different concentrations of sodium dodecyl sulfate. Moreover, different membranes were used to investigate the influence of surface chemistry. Finally, a standard oil-in-water emulsion was prepared and tested with MD. The composition simulates the composition of produced water, which is a common waste stream in the oil and gas production [29].

Appl. Sci. **2017**, *7*, 118

2. Materials and Methods

Three different microporous membranes were used in this study: PTFE (Tetratex, Donaldson Company Inc., Belguim, Leuven), polyethylene (PE) (Solupor®, Lydall Inc., Rochester, NH, USA) and oleophobic polyethersulfone (PES) (Supor®, Pall Corporation, New York, NY, USA). Sodium chloride (technical grade), Sodium dodecyl sulfate (SDS; 98%) and hexadecane (99%) and were purchased from Sigma Aldrich (Saint Louis, MO, USA).

Solutions with a concentration of 35 g/L sodium chloride and different concentrations of sodium dodecyl sulfate ranging from 10 to 150 mg/L were prepared under continuous stirring for 1 h. The oil-in-water emulsion was prepared with 2400 mg/L hexadecane, 240 mg/L sodium dodecyl sulfate and 10 g/L NaCl. The components were mixed and ultrasound was applied for 30 min to stabilize the suspension. For the oil-in-water emulsion, the size of the oil droplets was measured by a particle size analyzer (Nanosight NS500, Malvern Instruments Ltd., Malvern, UK) and was found in the range of 0.1–0.5 μm with an average particle diameter of 0.3 μm [29].

The surface tension of the liquids was measured using a Force Tensiometer K6 from Kruss GmbH (Hamburg, Germany). The contact angle of the membranes was measured with an OCA 15EC Contact Angle System of Dataphysics (Filderstadt, Germany) using the static sessile drop method. The critical surface tension of wetting was determined for the PES membrane by measuring the contact angles using a series of alkanes (from hexadecane to hexane). The surface tension of the liquid that first shows a contact angle below 90°, i.e., the liquid that wets the membrane is assumed as the critical surface tension. For PTFE and PE, literature values are used. The liquid entry pressure was determined as described by Khayet et al. [30]. The hydrostatic pressure was increased slowly by 0.1 bar each 30 s, until a flow was detected. The porosity was measured using helium pycnometry as described in [31]. The thickness was obtained by imaging the membrane cross-section using a cold field emission scanning electron microscope (SEM) type JSM6340F (JEOL, Tokyo, Japan) as described in [31]. The average and maximum pore size were measured using a Porolux® 1000, with Porefil as wetting liquid and the shape factor assumed to be 1 [31].

The direct contact membrane distillation (DCMD) experiments were carried out using the experimental setup described in [32]. The process scheme of the setup is visualized in Figure 1. The module had a membrane area of 0.0108 m^2 and 2 mm thick polypropylene (PP) spacers were used. The feed and permeate temperatures were kept constant at 60 and 45 °C, respectively, for each experiment. The flow velocity was 0.13 m/s at feed and permeate side. The first MD test included a stepwise increase of the concentration of SDS up to 150 mg/L. These experiments were run for 2 h at each concentration. Additionally, longer tests were performed using 150 mg/L SDS, where the flux was measured over three days.

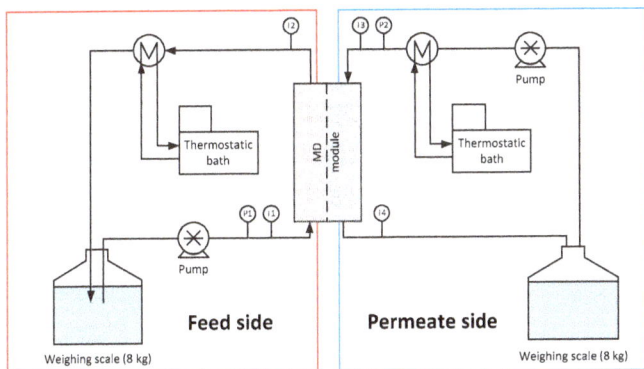

Figure 1. Process scheme of the direct contact membrane distillation (DCMD)-lab scale setup. T indicates a temperature sensor, P refers to pressure sensors.

3. Results

3.1. Membrane Characterization

The thickness (δ), porosity (ε) andmean pore size (d_{mean}) of the membranes used in this study is given in Table 1. These properties mainly determine the absolute flux [33]. The maximum pore size (d_{max}) and water contact angle (θ_{water}) are mostly important for the wetting resistance, which is sufficiently high for pure water. The water contact angle for the PTFE membrane is the highest (138°), followed by the PES membrane (132°) and the PE membrane (120°). The hexadecane contact angle ($\theta_{hexadecane}$) is used to test the oleophobicity of the membranes. As expected, PTFE and PE do not show any resistance to wetting of hexadecane, whereas the oleophobic PES membrane shows a contact angle of 92° with hexadecane.

Table 1. Properties of the membranes used in this study.

Membrane	δ (µm)	ε (%)	d_{mean} (µm)	d_{max} (µm)	θ_{water} (°)	$\theta_{hexadecane}$ (°)	LEP (bar)
PE	99	76%	0.30	0.43	138	0	3.9
PES	81	58%	0.51	0.59	132	92	4
PTFE	77	83%	0.17	0.19	120	0	10.8

3.2. Prediction of Membrane Wetting with Sodium Dodecyl Sulfate

3.2.1. Surface Tension

The surface tension of the aqueous solution of 35 g/L sodium chloride decreases with increasing concentrations of sodium dodecyl sulfate (SDS) (Figure 2). As described in the literature, the surface tension decreases strongly with increasing concentration of SDS up to 50 mg/L, after which an asymptotic level is reached at higher concentrations of sodium dodecyl sulfate [33–35]. This point correlates with the critical micelle concentration (CMC) of the surfactant. For pure water, a the critical micelle concentration of 2560 g/L is reported for sodium dodecyl sulfate systems, whereas it is also observed that for salt-containing systems this value is much lower [35]. This means that salinity is an important aspect when studying the wetting behavior of a solution containing surfactants.

The critical surface tension for wetting, defined as the surface tension required to wet the membrane, is also indicated in the figure. Based on this figure, the PE membrane is expected to be wetted by the feed liquid if it contains 50 mg/L SDS or more. The surface energy of the PTFE and PES membranes is below the asymptotic value of the surface tension of the liquid up to 150 mg/L and therefore these membranes might be more suitable for treatment of surfactant containing waste streams.

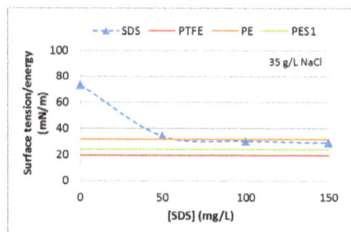

Figure 2. Surface tension as function of sodium dodecyl sulfate (SDS) concentration in an aqueous solution of 35 g/L NaCl and critical surface tension for the membrane surface used in this study.

3.2.2. Contact Angle

Figure 3 shows the contact angle of the three different membranes as a function of the sodium dodecyl sulfate concentration. The PTFE and PES membrane show contact angles above 90° for all SDS concentrations used in this study, whereas the contact angle with PE decreases below 90° for SDS concentrations above 50 mg/L. Based on these results, wetting is only expected for the PE membrane at SDS concentrations higher than 50 mg/L, which confirms the observations in Section 3.2.1.

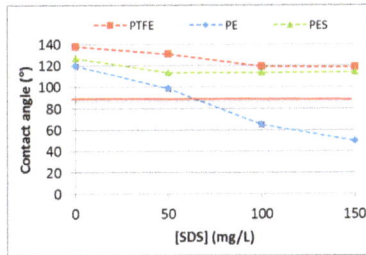

Figure 3. Contact angle measurements at different SDS concentrations, [NaCl] = 35 g/L.

3.2.3. Liquid Entry Pressure

Figure 4 presents the LEP for the three different membranes for a variety of SDS concentrations and 35 g/L NaCl. Up to a concentration of 50 mg/L SDS no immediate breakthrough of the liquid was observed for a pressure up to 4 bar for all three membranes. The maximum pressure of the experimental setup was 4 bar. Higher liquid entry pressures are therefore not measurable, which is indicated by the hatched bars. The PE and PES membrane shows a strong decline of LEP at higher SDS concentrations. However, in contrast to what was expected based on the surface tension and contact angle measurements, an LEP of 1.8 bar was still achieved at 150 mg/L SDS for the PE membrane. Neither the PES nor the PTFE membrane showed a severe decrease in contact angle and surface tension. Nevertheless, the lowest LEP of 1.4 bar at 150 mg/L SDS is observed for the PES membrane, while the PTFE membrane still shows a liquid entry pressure of 3.5 bar. This difference in behavior of the LEP is not expected based on the contact angle measurements, but can be explained due to a difference in pore size (Table 1), which is inversely correlated to liquid entry pressure (Equation (1)). The PES membrane has the highest maximum pore diameter of 0.59 μm, while the PTFE membrane has a maximum pore diameter of only 0.19 μm.

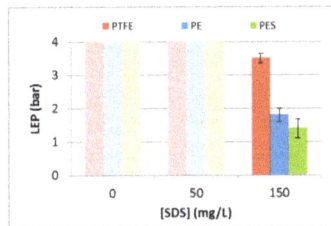

Figure 4. Liquid entry pressure at different SDS concentrations, [NaCl] = 35 g/L.

To measure the time dependence of the wetting an additional experiment was carried out. The membrane was placed inside the pressure cell and a constant pressure of 1 bar was applied, using a feed liquid of 150 mg/L SDS and 35 g/L NaCl. The pressure of 1 bar was selected based on the expected pressure drops in a full-scale MD module [36,37]. The PTFE membrane does not show any liquid penetrating the membrane after 30 min, whereas the PE and PES membranes show liquid breakthrough after 11 and 18 min, respectively. This shows that the liquid entry pressure measurements in Figure 4, with a pressure step of 0.1 bar each 30 s are not representative for long-term membrane operation in membrane distillation. The differences between the MD process and the liquid entry pressure tests might explain this behavior. A few wetted pores are only detected after sufficient time in the liquid entry measurement, while in MD an immediate increase of the permeate conductivity is observed. In this respect, membrane distillation is much more sensitive in detection of leakages. The prediction of membrane wetting based on liquid entry pressure using this procedure might therefore be an underestimation and it is recommended that the time dependence must carefully be considered when evaluating the wetting behavior of a membrane.

3.2.4. Wetting Prediction

Based on surface tension and the water contact angle, no wetting is expected for the PTFE and PES membrane. The surface tension with a concentration of SDS of 50 mg/L decreases below the critical surface energy of PE. This is also visualized by the water contact angle, which decreases below 90° for concentrations above 50 mg/L SDS. Based on these two techniques, wetting in the MD tests is expected at a concentration of 50 mg/L SDS. In contrast, the liquid entry pressure for all membranes is above 1.4 bar, while at lab scale a pressure drop of 20 mbar is expected. This indicates that based on liquid entry measurement, no wetting is expected for the lab-scale MD tests. However, at a larger scale, MD pressure drop might increase up to 1 bar. The observation of leakage after applying a hydrostatic pressure difference of 1 bar on the membrane using 150 mg/L SDS points to possible membrane wetting in full scale modules for the PE and PES membrane.

3.3. Membrane Distillation with SDS

3.3.1. PE

Figure 5 shows the average flux and salt retention of the PE membrane using different SDS concentrations. Up to 40 mg/L, the flux is unchanged and the salt retention remains sufficiently high (>99.9%). At 50 mg/L SDS, the flux increases steadily over time. During the first hour of the experiment at 50 mg/L SDS, the flux was relatively stable, at 17 $kg\cdot h^{-1}\cdot m^{-2}$, while the retention already decreased from 99.9% to 98.6%. These observations indicate wetting of a few membrane pores, enabling limited salt transport through the membrane. The fraction of dry membrane pores must be substantially high, because the flux is not affected and the salt retention remains above 98.6%. During the second hour of the experiment, the flux increased rapidly from 17 $kg\cdot h^{-1}\cdot m^{-2}$ up to 40 $kg\cdot h^{-1}\cdot m^{-2}$, while the retention decreased drastically from >98.6% to 69.5%, which indicates severe wetting of the PE membrane at these concentrations. The SDS concentration is not further increased above 50 mg/L, because even more severe wetting problems are expected. Based on the contact angle measurements, wetting would occur between 50 and 100 mg/L, while based on the LEP measurement, no wetting is expected up to 1000 mg/L. This shows that the contact angle and surface tension are more reliable measures for prediction of membrane wetting than the LEP. The short-term exposure LEP in Section 3.2.3 did not predict membrane wetting under these experimental condition, and is therefore found to be an unreliable measure for wetting prediction.

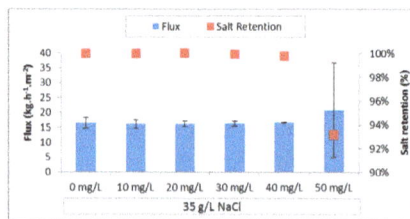

Figure 5. Flux and salt retention for the PE membrane for a feed composition with different SDS concentrations and 35 g/L NaCl.

3.3.2. PTFE

The same concentrations of SDS were used with the PTFE membrane. The fluxes and salt retentions for each concentration are given in Figure 6a. The PTFE membrane shows a stable performance over 2 h of operation up to an SDS concentration of 150 mg/L, as is expected based on the water contact angle and LEP measurements. Figure 6b shows the daily average flux and retention for the MD experiment with 150 mg/L SDS and 35 g/L NaCl. Despite the fact that a stable performance was observed for the short-term experiments over 2 h, longer-term experiments show the intolerance of PTFE membranes for surfactants. Partial pore wetting is gradually provoked by the surfactant and results in salt transport from feed to permeate.

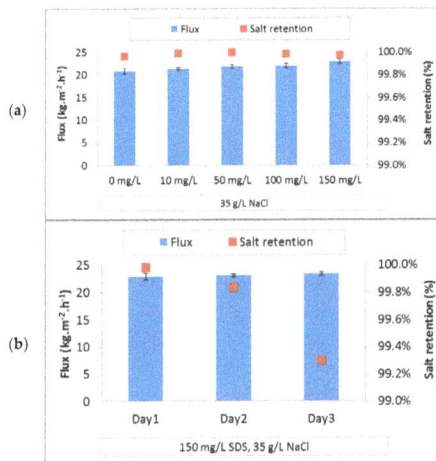

Figure 6. Flux and salt retention for the PTFE membrane; (**a**) with increasing SDS concentrations; (**b**) At constant feed composition over three days.

3.3.3. PES

In contrast to the PE and PTFE membrane, the flux and retention of the PES membrane did not indicate any wetting at the different SDS concentrations used in this study (Figure 7). In addition, a three-day test at the highest SDS concentration of 150 mg/L shows salt retention. This behavior is in accordance with the predictions based on the surface tension and the contact angle measurement. The liquid entry pressure of 1.4 bar at 150 mg/L SDS is sufficient for lab-scale testing, because in the lab setup, the pressure drop did not exceed 0.02 bar for the process conditions used in this study. Nevertheless, it is remarkable that the PES membrane showed a lower LEP compared to the PTFE membrane, while it does show a better salt retention during the MD-tests. As observed during the determination of the critical surface tension, the PES membrane does not wet using hexadecane, proving its oleophobic character. These tests show that this membrane feature can improve the performance of the membrane in terms of wetting resistance in the presence of surface-lowering components.

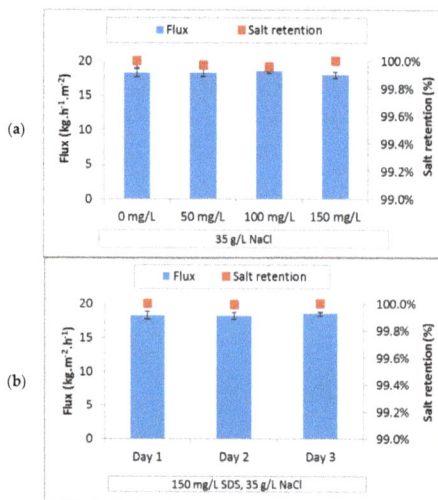

Figure 7. Flux and salt retention for the PES membrane, (**a**) with increasing SDS concentrations; (**b**) at constant feed composition over three days.

3.3.4. Summary

The best prediction for the immediate wetting in membrane distillation was obtained using the surface tension measurement combined with the contact angle of the fluid with the membrane. Based on this quick and easy measurement, a first estimation of the technical feasibility and a selection of the membrane can be made. However, long-term gradual wetting, as observed with the PTFE membrane, was not predicted by surface tension, contact angle or liquid entry pressure. In addition, when applying the liquid at 1 bar for a longer time (30 min) no wetting was observed, indicating that the long-term membrane wetting as observed in the PTFE membrane is more difficult to predict using quick analytic experiments.

3.4. Synthetic Produced Water

3.4.1. Prediction of Membrane Wetting

An emulsion of hexadecane in water was prepared to study the resistance of membrane against wetting in the presence of oily substances. The surface tension of the synthetic produced water was 47 mN/m, which is much higher compared to the lowest surface tension of the solutions with sodium dodecyl sulfate of ± 32 mN/m (Figure 2). It is expected that the sodium dodecyl sulfate forms micelles around the hexadecane, shielding the hexadecane from the air–liquid interface. The oleophobic tail emulsifies the oil, at the same time diminishing its capability of reducing the interfacial air-water surface tension (Figure 8). The contact angle of the synthetic produced water with PTFE and PES was 127°. Membrane wetting was therefore not expected for either membrane. Due to the severe wetting of the PE membrane observed during the SDS experiments, this membrane was not further considered for the treatment of oil emulsions.

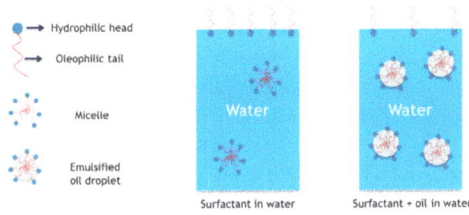

Figure 8. Micelle formation with surfactant in water mixture and emulsification of the oil-in-water mixture.

3.4.2. Membrane Distillation Testing

Figure 9 shows the flux of the PTFE and the PES membrane during a one-day test. The flux using the PTFE membrane was stable up to 4 h of operation. At that moment, water was added to the feed to keep the feed concentration constant. Thereafter, the flux decreases to zero. Figure 10 shows the visual transformation from opaque to transparent of the PTFE membrane, which shows the gradual wetting during the course of the experiment. Remarkably, no increase was observed in permeate conductivity for the PTFE, which remained below 20 μS/cm. When refilling the feed vessel with water, the emulsion is locally broken (observed visually) and hexadecane is able to penetrate the membrane. Since the hexadecane and PTFE are both apolar molecules, the membrane takes up the hexadecane, which remains in the membrane and blocks the flux. Sodium chloride is not soluble in hexadecane, explaining the unexpected combination of flux decrease due to membrane wetting, without loss of salt retention. In contrast to PTFE, the flux of the PES membrane only steadily decreases, with only 9%. No visual observation of wetting was observed for this membrane, however the steady decrease might indicate that some of the pores are also blocked by the hexadecane, reducing the flux, causing the slight decrease in flux.

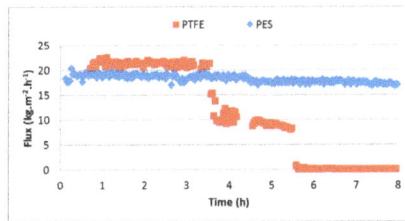

Figure 9. Flux as function of time for PTFE and PES with the oil/SDS mixture as feed solution.

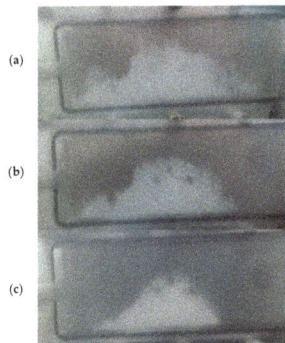

Figure 10. Pictures of the PTFE membrane during the experiment with oil/SDS mixture after (**a**) 4 h; (**b**) 5 h; (**c**) 6 h of MD operation.

4. Conclusions

In this manuscript it was shown that immediate wetting in membrane distillation can be predicted by surface tension and water contact angle measurements. The surface tension measurement shows that wetting behavior might not only depend on sodium dodecyl concentration, but also on the NaCl concentration, which strongly affects the critical micelle concentration. The PE membrane is less hydrophobic compared to PTFE and is therefore more susceptible towards membrane wetting due to the presence of surfactants. This membrane shows immediate wetting, a severe salt increase in the permeate, and a flux increase induced by the hydrostatic pressure difference. For PTFE the short-term test did not show changes in salt retention for SDS concentrations far above the critical micelle concentration (SDS: 150 mg/L, NaCl: 35 g/L). However, three-day MD testing shows that despite the sufficiently high contact angle with the PTFE-membrane, a decrease of the salt retention is observed, while still maintaining stable fluxes. This shows that the wetting is much less severe compared to the PE membrane, but that at longer operational times membrane wetting might also become an important issue. The PES membrane is oleophobic and showed unaffected flux or salt retention over three days of testing. Additionally, membrane distillation experiments with synthetic produced water showed the better stability in performance of an oleophobic membrane compared to a hydrophobic PTFE membrane.

Acknowledgments: Lies Eykens thankfully acknowledges a PhD scholarship provided by the Flemish Institute for Technological Research, VITO NV.

Author Contributions: Lies Eykens performed and analyzed the experiments and wrote the paper. All other authors contributed significantly through their guidance and support during the experiments and the writing process.

Conflicts of Interest: The authors declare no conflict of interest.

References

1. Lin, S.; Yip, N.Y.; Elimelech, M. Direct contact membrane distillation with heat recovery: Thermodynamic insights from module scale modeling. *J. Membr. Sci.* **2014**, *453*, 498–515. [CrossRef]
2. Kesieme, U.K.; Milne, N.; Aral, H.; Cheng, C.Y.; Duke, M. Economic analysis of desalination technologies in the context of carbon pricing, and opportunities for membrane distillation. *Desalination* **2013**, *323*, 66–74. [CrossRef]
3. Sha, D.L.; Chavez, L.H.A.; Ben-sasson, M.; Castrillo, S.R. Desalination and Reuse of High-Salinity Shale Gas Produced Water: Drivers, Technologies, and Future Directions. *Environ. Sci. Technol.* **2013**, *47*, 9569–9583.
4. Alkhudhiri, A.; Darwish, N.; Hilal, N. Treatment of high salinity solutions: Application of air gap membrane distillation. *Desalination* **2012**, *287*, 55–60. [CrossRef]
5. Duong, H.C.; Chivas, A.R.; Nelemans, B.; Duke, M.; Gray, S.; Cath, T.Y.; Nghiem, L.D. Treatment of RO brine from CSG produced water by spiral-wound air gap membrane distillation—A pilot study. *Desalination* **2015**, *366*, 121–129. [CrossRef]
6. Ge, J.; Peng, Y.; Li, Z.; Chen, P.; Wang, S. Membrane fouling and wetting in a DCMD process for RO brine concentration. *Desalination* **2014**, *344*, 97–107. [CrossRef]
7. Martinetti, C.R.; Childress, A.E.; Cath, T.Y. High recovery of concentrated RO brines using forward osmosis and membrane distillation. *J. Membr. Sci.* **2009**, *331*, 31–39. [CrossRef]
8. Criscuoli, A.; Zhong, J.; Figoli, A.; Carnevale, M.C.; Huang, R.; Drioli, E. Treatment of dye solutions by vacuum membrane distillation. *Water Res.* **2008**, *42*, 5031–5037. [CrossRef] [PubMed]
9. El-Abbassi, A.; Hafidi, A.; Khayet, M.; García-Payo, M.C. Integrated direct contact membrane distillation for olive mill wastewater treatment. *Desalination* **2013**, *323*, 31–38. [CrossRef]
10. Prince, J.A.; Singh, G.; Rana, D.; Matsuura, T.; Anbharasi, V.; Shanmugasundaram, T.S. Preparation and characterization of highly hydrophobic poly(vinylidene fluoride)-Clay nanocomposite nanofiber membranes (PVDF-clay NNMs) for desalination using direct contact membrane distillation. *J. Membr. Sci.* **2012**, *397–398*, 80–86. [CrossRef]
11. Thiel, G.P.; Tow, E.W.; Banchik, L.D.; Chung, H.W.; Lienhard, J.H. Energy consumption in desalinating produced water from shale oil and gas extraction. *Desalination* **2015**, *366*, 94–112. [CrossRef]

12. Warsinger, D.M.; Swaminathan, J.; Guillen-Burrieza, E.; Arafat, H.A.; Lienhard, J.H. Scaling and fouling in membrane distillation for desalination applications: A review. *Desalination* **2015**, *356*, 294–313. [CrossRef]

13. Hausmann, A.; Sanciolo, P.; Vasiljevic, T.; Weeks, M.; Schroën, K.; Gray, S.; Duke, M. Fouling mechanisms of dairy streams during membrane distillation. *J. Membr. Sci.* **2013**, *441*, 102–111. [CrossRef]

14. Tijing, L.D.; Woo, Y.C.; Choi, J.S.; Lee, S.; Kim, S.H.; Shon, H.K. Fouling and its control in membrane distillation—A review. *J. Membr. Sci.* **2015**, *475*, 215–244. [CrossRef]

15. Hausmann, A.; Sanciolo, P.; Vasiljevic, T.; Weeks, M.; Schroën, K.; Gray, S.; Duke, M. Fouling of dairy components on hydrophobic polytetrafluoroethylene (PTFE) membranes for membrane distillation. *J. Membr. Sci.* **2013**, *442*, 149–159. [CrossRef]

16. Nguyen, Q.M.; Lee, S. Fouling analysis and control in a DCMD process for SWRO brine. *Desalination* **2015**, *367*, 21–27. [CrossRef]

17. Guillen-Burrieza, E.; Ruiz-Aguirre, A.; Zaragoza, G.; Arafat, H.A. Membrane fouling and cleaning in long term plant-scale membrane distillation operations. *J. Membr. Sci.* **2014**, *468*, 360–372. [CrossRef]

18. Curcio, E.; Ji, X.; Di Profio, G.; Sulaiman, A.O.; Fontananova, E.; Drioli, E. Membrane distillation operated at high seawater concentration factors: Role of the membrane on $CaCO_3$ scaling in presence of humic acid. *J. Membr. Sci.* **2010**, *346*, 263–269. [CrossRef]

19. Nghiem, L.D.; Cath, T. A scaling mitigation approach during direct contact membrane distillation. *Sep. Purif. Technol.* **2011**, *80*, 315–322. [CrossRef]

20. Garcia-Payo, M.M.C.; Izquierdo-Gil, M.A.M.; Fernandez-Pineda, C. Wetting Study of Hydrophobic Membranes via Liquid Entry Pressure Measurements with Aqueous Alcohol Solutions. *J. Colloid Interface Sci.* **2000**, *230*, 420–431. [CrossRef] [PubMed]

21. Gryta, M.; Barancewicz, M. Influence of morphology of PVDF capillary membranes on the performance of direct contact membrane distillation. *J. Membr. Sci.* **2010**, *358*, 158–167. [CrossRef]

22. Gryta, M. Concentration of NaCl solution by membrane distillation integrated with crystallization. *Sep. Sci. Technol.* **2002**, *37*, 3535–3558. [CrossRef]

23. Bilad, M.R.; Guillen-Burrieza, E.; Mavukkandy, M.O.; Al Marzooqi, F.A.; Arafat, H.A. Shrinkage, defect and membrane distillation performance of composite PVDF membranes. *Desalination* **2015**, *376*, 62–72. [CrossRef]

24. Lalia, B.S.; Guillen-Burrieza, E.; Arafat, H.A.; Hashaikeh, R. Fabrication and characterization of polyvinylidenefluoride-co-hexafluoropropylene (PVDF-HFP) electrospun membranes for direct contact membrane distillation. *J. Membr. Sci.* **2013**, *428*, 104–115. [CrossRef]

25. Franken, A.C.M.; Nolten, J.A.M.; Mulder, M.H.V.; Bargeman, D.; Smolders, C.A. Wetting criteria for the applicability of membrane distillation. *J. Membr. Sci.* **1987**, *33*, 315–328. [CrossRef]

26. Lin, S.; Nejati, S.; Boo, C.; Hu, Y.; Osuji, C.O.; Elimelech, M. Omniphobic Membrane for Robust Membrane Distillation. *Environ. Sci. Technol. Lett.* **2014**, *1*, 443–447. [CrossRef]

27. Wang, Z.; Hou, D.; Lin, S. Composite Membrane with Underwater-Oleophobic Surface for Anti-Oil-Fouling Membrane Distillation. *Environ. Sci. Technol.* **2016**, *50*, 3866–3874. [CrossRef] [PubMed]

28. Zuo, G.; Wang, R. Novel membrane surface modification to enhance anti-oil fouling property for membrane distillation application. *J. Membr. Sci.* **2013**, *447*, 26–35. [CrossRef]

29. Mustafa, G.; Wyns, K.; Buekenhoudt, A.; Meynen, V. Antifouling grafting of ceramic membranes validated in a variety of challenging wastewaters. *Water Res.* **2016**, *104*, 242–253. [CrossRef] [PubMed]

30. Khayet, M.; Matsuura, T. Preparation and Characterization of Polyvinylidene Fluoride Membranes for Membrane Distillation. *Ind. Eng. Chem. Res.* **2001**, *40*, 5710–5718. [CrossRef]

31. Eykens, L.; De Sitter, K.; Dotremont, C.; Pinoy, L.; Van der Bruggen, B. Characterization and performance evaluation of commercially available hydrophobic membranes for direct contact membrane distillation. *Desalination* **2016**, *392*, 63–73. [CrossRef]

32. Eykens, L.; Hitsov, I.; De Sitter, K.; Dotremont, C.; Pinoy, L.; Nopens, I.; Van der Bruggen, B. Influence of membrane thickness and process conditions on direct contact membrane distillation at different salinities. *J. Membr. Sci.* **2016**, *498*, 353–364. [CrossRef]

33. Hernáinz-Bermúdez de Castro, F.; Gálvez-Borrego, A.; Calero-de Hoces, M. Surface Tension of Aqueous Solutions of Sodium Dodecyl Sulfate from 20 °C to 50 °C and pH between 4 and 12. *J. Chem. Eng. Data* **1998**, *43*, 717–718. [CrossRef]

34. Kloubek, J. Measurement of the dynamic surface tension by the maximum bubble pressure method. IV. Surface tension of aqueous solutions of sodium dodecyl sulfate. *J. Colloid Interface Sci.* **1972**, *41*, 17–22. [CrossRef]

35. Owens, D.K.D. The dynamic surface tension of sodium dodecyl sulfate solutions. *J. Colloid Interface Sci.* **1969**, *29*, 496–501. [CrossRef]

36. Hitsov, I.; Eykens, L.; De Schepper, W.; De Sitter, K.; Dotremont, C.; Nopens, I. Full-scale Direct Contact Membrane Distillation (DCMD) model including membrane compaction effects. *J. Membr. Sci.* **2017**, *524*, 245–256. [CrossRef]

37. Winter, D.; Koschikowski, J.; Wieghaus, M. Desalination using membrane distillation: Experimental studies on full scale spiral wound modules. *J. Membr. Sci.* **2011**, *375*, 104–112. [CrossRef]

applied
sciences

MDPI

Article

Exergy Analysis of Air-Gap Membrane Distillation Systems for Water Purification Applications

Daniel Woldemariam [1,2,*], Andrew Martin [1] and Massimo Santarelli [1,2]

[1] Energy Technology Department, KTH Royal Institute of Technology, Brinellvägen 68,
SE-100 44 Stockholm, Sweden; andrew.martin@energy.kth.se (A.M.); massimo.santarelli@polito.it (M.S.)

[2] Department of Energy (DENERG), Politecnico di Torino, PoliTo, Corso Duca degli Abruzzi 24,
10129 Turin, Italy

* Correspondence: dmwo@kth.se; Tel.: +46-08-790-7477

Academic Editor: Enrico Drioli
Received: 13 December 2016; Accepted: 15 March 2017; Published: 20 March 2017

Abstract: Exergy analyses are essential tools for the performance evaluation of water desalination and other separation systems, including those featuring membrane distillation (MD). One of the challenges in the commercialization of MD technologies is its substantial heat demand, especially for large scale applications. Identifying such heat flows in the system plays a crucial role in pinpointing the heat loss and thermal integration potential by the help of exergy analysis. This study presents an exergetic evaluation of air-gap membrane distillation (AGMD) systems at a laboratory and pilot scale. A series of experiments were conducted to obtain thermodynamic data for the water streams included in the calculations. Exergy efficiency and destruction for two different types of flat-plate AGMD were analyzed for a range of feed and coolant temperatures. The bench scale AGMD system incorporating condensation plate with more favorable heat conductivity contributed to improved performance parameters including permeate flux, specific heat demand, and exergy efficiency. For both types of AGMD systems, the contributions of the major components involved in exergy destruction were identified. The result suggested that the MD modules caused the highest fraction of destructions followed by re-concentrating tanks.

Keywords: exergy; energy; membrane distillation; specific heat; entropy; efficiency

1. Introduction

Exergy studies on membrane-based desalination and water purification technologies such as reverse osmosis (RO) and membrane distillation (MD) focus on evaluating the exergetic efficiencies of the components, mainly membrane modules. Unlike energy, exergy is destroyed and can only be conserved when all processes occurring in a system and its surrounding environment are reversible [1]. This thermodynamic irreversibility in a system can be quantified and referred to as exergy destruction. Therefore, the exergy efficiency of processes is a measure of their approach to ideality or reversibility [2]. The exergy rates in streams of processes associated with heat transfer such as MD depend mainly on the temperature at which the process occurs in relation to the temperature of the environment. The exergy efficiencies of such processes are dependent on heat recovered and heat losses through module surfaces to the surrounding atmosphere. Hence, exergy analysis provides unique insights into the types, locations, and causes of losses and aids in identifying improved thermal integration.

Membrane distillation involves the separation of water from a preheated feed by subjecting it to phase change and the subsequent passage through a hydrophobic porous membrane to yield pure water upon condensation. Membrane distillation requires heat to evaporate the feed and electricity for the low pressure pumps. The heat sources used in MD include solar [3–7], district heat [8], or waste heat [9,10]. Though low-grade heat is sufficient to drive the MD process, thermal energy demand is high, and this together with low permeate flux create challenges for the technology. The high heat demand is largely related to the substantial amount of heat transferred through the membrane, mainly as latent heat but partially also from conductive heat transfer. However, the ability to supply an ultra high-quality permeate together with the possibility to drive the MD process with low grade or waste heat are the opportunities gaining acceptance in large scale industrial applications.

Exergy analyses of membrane-based desalination or water purification processes have been investigated by few researchers. Banat and Jwaied [4] conducted exergy destruction analyses for compact and large-scale solar-powered MD systems. They reported that most of the exergy destruction was in the MD modules with 98.8% and 55.14% for compact and large scale systems, respectively. Exergy efficiency of a 24,000 m^3/day DCMD desalination plant operated with and without heat recovery system was reported as 28.3% and 25.6% for the desalination plant with and without heat recovery, respectively [11]. They also reported energy consumptions of 39.7 kWh/m^3 and 45 kWh/m^3 for the same plant with and without heat recovery, respectively.

Another study [12] on energetic and exergetic analysis of RO/MD and NF/RO/MD hybrid systems for production capacities of 904 and 836 m^3/h, respectively, reported specific thermal energy demand ranging from 2.25 to 15 kWh/m^3. The exergy destructions due to entropy production for each hybrid system was reported in the range of 7300 to 15,100 MJ/h. Macedonio and Drioli carried out an energetic, exergetic, and economic evaluation of seawater desalination [13]. The integrated system consisted of microfiltration, nanofiltration, membrane crystallization (MCr), RO, and MD, where MCr was applied on NF retentate, and MD was applied on RO retentate. Without energy recovery, the specific energy demand (SED) was found to be 28 kWh/m^3. When a pressure exchanger is included as an energy recovery device, a slightly lower SED (27.5 kWh/m^3) was obtained. The reported exergetic efficiencies ranged from 15.8% without energy recovery to 21.9% when including the pressure exchanger and internal heat recovery. Sow et al. [14] analyzed solar-driven distillers and reported exergy efficiencies of 19%–26%, 17%–20%, and less than 4% for triple-effect, double-effect, and single-effect systems. Garcia-Rodriguez and Gomez-Cmacho [15] analyzed exergetic efficiencies of a 14-cell multi-effect distillation (SOL-14 MED) plant in Plataforma Solar de Almeria, Spain. They developed a thermodynamic computer program in order to propose improvements. The reported results showed that exergy efficiency of the plant increases from 1.4% to 4.7% and from 14.3% to 25.7% because of the integration of a double-effect absorption heat pump and energy recovery, respectively.

Thermal energy demands have been analyzed for membrane-based water desalination systems. Solar powered MD was reported to have a specific heat demand of 140–200 kWh/m^3 [6], whereas a solar thermal and PV integrating stand-alone MD plant was reported to have a heat demand of 200–300 kWh/m^3 [16]. Islam et al. analyzed exergy destructions in different components of a two-stage 553 m^3/h capacity reverse osmosis unit in a thermal power plant. The reported contributions for exergy destructions showed a maximum exergy loss in throttling valves (57%) followed by RO modules (21%), with plant's overall second-law efficiency of 4.1% [17]. Energy (first-law) analysis is widely used for assessing efficiencies of water desalination systems. However, as it does not account for the quality of energy considered, second-law analysis is required to decide how much of the available energy is destroyed, for example, in heat exchangers and MD modules. Exergy analysis also helps to identify components responsible for deterioration of the available energy in terms of exergy destruction and hence helps to suggest unit redesign. It is clear that MD modules are responsible for the majority of the exergy destruction in water purification or desalination systems employing this technology. The challenges to be achieved in this study that have not been addressed previously are summarized in the following questions: What is the exergy destruction contribution of each component of AGMD system? How would temperature difference across the module affect the exergy efficiency of the modules? What would the applications for water purification be It was believed that the answers to these questions would be useful for developing or redesigning large-scale MD modules with improved exergy performances. Indeed, temperature dependency on exergy efficiency and exergy destruction is a key parameter, since MD technology offers freedom in selecting heat source and heat sink temperatures when integrated with other processes. Such aspects have not been covered in sufficient detail in previous studies. Moreover, to the authors' knowledge, air gap membrane distillation (AGMD) has not been previously considered in this context.

The main objective of this study was to evaluate the exergy efficiencies and destruction of AGMD systems. The analyses are based on two types of semi-commercial AGMD modules for a new application of water treatment where contaminants in the feed are at relatively low concentrations. The study covers calculations of the exergy destruction contributions of the components of each MD system considered. Additionally, effects of feed–coolant temperature differences are analyzed, and comparisons are made with other studies on similar and related water purification systems.

2. Materials and Methods

Thermodynamic properties of streams used for the calculation of exergy analyses were obtained from testing two AGMD modules, Xzero and Elixir500 at IVL Hammarby Sjöstadsverk and KTH, respectively. Both modules have identical membrane materials of PTFE with PP backing but differ in the total membrane area and the nature of the condensation surface. The Xzero unit was originally intended for industrial applications such as semiconductor fabrication facilities (fabs), whereas the Elixir500 unit was designed for small-scale water purification. The Xzero module utilizes PP plates and the Elixir500 module employs a thin plate of stainless steel condensation surface, the same design as plate-and-frame heat exchangers. Both MD modules are assembled and supplied by Scarab Development AB. More information on these modules can be found in the literature [8,18].

2.1. Experimental Setup for the Pilot AGMD System

The pilot MD system analyzed in this study consists of 10 AGMD modules arranged in five cascades, each containing two modules connected in series. Figure 1 shows the photograph of the pilot MD system. The AGMD system is connected to the district heating network through heat exchangers, which supply the heat required to heat up the MD feed. The major components include the AGMD modules, the recirculation tank with a 1 m^3 capacity, and an expansion tank for pressure balancing on the coolant side. Two plate and frame heat exchangers (nominal capacity of 300 kW, working temp. of 0–170 °C) and two pumps (each 2.2 kW power demand at Q_{max} = 17 m^3/h and H_{max} 28.7 m pressure head) are also connected.

Figure 1. Xzero Pilot air gap membrane distillation system at Hammarby Sjöstadsverk testing and demonstration facility, Stockholm, Sweden.

An illustration of the pilot plant depicting all the streams involved is shown in Figure 2. Each MD module is constructed from a stack of 10 cassettes, each containing two membranes with a combined area of 0.28 m². The membrane characteristics are a 0.2 mm thickness, a 0.2 μm average pore size, and an 80% porosity. The size of one module is 63 cm wide and 73 cm high, with a stack thickness of 17.5 cm. During the experiments, data was logged for parameters connected to the control system. PT100 thermocouples (range: −50 °C–200 °C, accuracy: ±4 °C) installed close to the MD module's inlet and outlet streams were used to measure temperatures. FLEX-F flow sensors (linear flow range: 2–150 cm/s, accuracy: ±10 cm/s) were used to measure the flow rates of streams, which were recorded by the control system. System control and data (temperature, pressure, and the flow rates of the feed and cooling water) were registered on a personal computer with Citect Runtime SCADA software installed. A YOKOGAWA DC402G dual cell conductivity analyzer (ranges of detection 1 μS/cm–25 mS/cm and accuracy of ±0.5%) was introduced to monitor the conductivity of the product water.

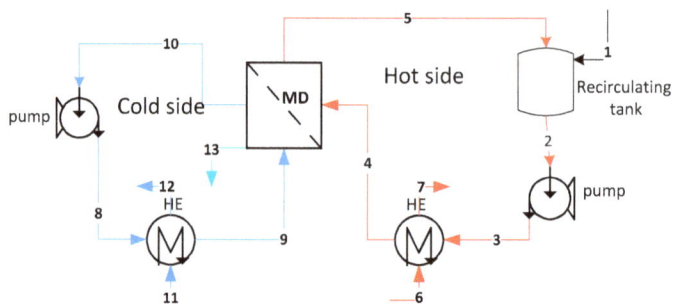

Figure 2. The Xzero AGMD unit used and the different streams considered.

2.2. Experimental Setup for Bench Scale AGMD

The bench scale AGMD module includes a thin plate of stainless steel inserted as a condensation plate, an effective membrane area of 0.19 m², with an 80% porosity and a 0.2 mm thickness. Figure 3 shows a photograph of the Elixir500 lab unit tested at KTH.

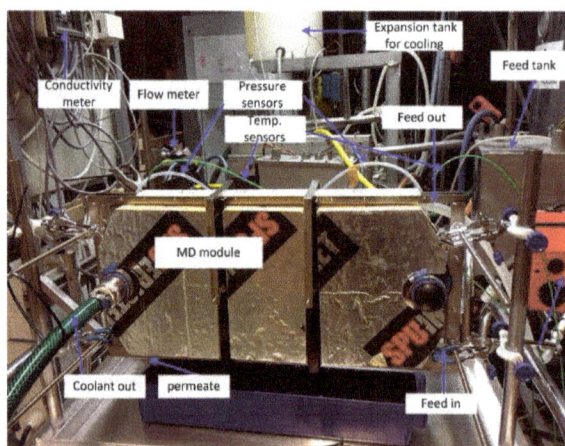

Figure 3. Bench scale Elixir500 AGMD test unit at KTH Royal Institute of Technology, Energy Department laboratory.

In Figure 4, the schematic diagram shows the main components and water streams (1–8) considered during analyses. The red lines show streams on the hot feed side of the module and the blue lines for the streams on the cooling side. Stream 8 refers to the permeate collected from the module. The bench scale MD unit consists of one MD unit module, a 20 L capacity heating, a recirculating tank, two pumps for each of the coolant, and feed streams. Two heaters immersed in the tank with a combined heating rate of 4.5 kW provide temperature control for the feed water.

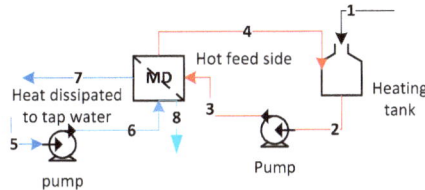

Figure 4. Water and heat streams considered in the Elixir500 AGMD lab unit.

The obtained permeate was returned to the feed tank so that feed concentration would be constant over the course of experimentation. In order to measure the feed and cold-side temperatures, thermocouples were installed at the inlets and outlets of the module and were connected to a data logger (Keithley 2701 DMM with a 7706 card). Once-through tap water was used as a heat sink, which exchanges heat with an external water to vary coolant temperature. Rotameters with built-in control valves measured feed and coolant water flow rates. Product water was measured with a graduated cylinder and stopwatch, typically during a 30 min period of steady operation.

The reference state and experimental conditions are given in Table 1. Data for the exergy analysis is taken for one cascade of MD modules containing two modules connected in series. Pressure drops across the MD modules are very low—less than 0.019 MPa. The concentration of total dissolved solutes in the feed water is also relatively low in comparison to seawater desalination applications (two orders of magnitude or less). As the three different cases have different feed and cooling temperatures, the exergy rates will be mainly affected by them.

Table 1. The experimental and reference state conditions for the two types of AGMD systems tested.

Property	Value	
	Xzero	Elixir500
Reference temperature, K	288	
Reference pressure, MPa	0.1	
Membrane area, m^2	4.6	0.19
Feed rate, L/min	20	3.8
Coolant rate, L/min	20	1.9
Feed temperature, K	353	338–353
Coolant temperature, K	288–328	
TDS of feed and coolant, g/L	0.25	0.36

2.3. Exergy Analysis Methods

The exergy rates of each stream throughout the MD systems and the exergy efficiencies of the MD modules are analyzed by taking the basic exergy definitions [1] and relationships for MD models as reported in the literature [5,6,9]. The intensive parameters governing the MD process are temperature and composition, and exergies related to kinetic and potential are not considered. Hence, the exergy rate is the sum of the physical and chemical exergies of streams as given by Equation (1).

$$Ex = Ex_{ph} + Ex_{ch} \tag{1}$$

Ex_{ph} represents the thermomechanical exergy based on the temperature of the stream, and Ex_{Ch} refers to the chemical exergy from the chemical potentials of the solute components in the stream. The physical exergy can be expressed in terms of heat enthalpies and entropies at the specified condition (h, s) and reference condition (h_o, s_o) by Equation (2).

$$Ex_{ph} = (h - h_o) - T_o \times (s - s_o) \tag{2}$$

The individual exergies due to pressure and temperature are calculated independently and sum up to the physical exergy. Even though the working pressure in MD is generally very low, its effect could be detected in exergy destruction calculations due to pressure changes, especially in series configured modules. These pressure and temperature related exergies are given as follows, in Equations (3) and (4) [5,10].

$$Ex_p = \dot{m} \times (p - p_o)/\varrho \tag{3}$$

$$Ex_T = \dot{m} \times c_p \times [(h - h_o) - T_o(s - s_o)] \tag{4}$$

Chemical exergy based on the chemical potentials or concentrations (C) of the components in the stream is calculated as in Equation (5).

$$Ex_c = -\dot{m} \times N_s \times R \times T_o \times \ln X_s \tag{5}$$

$$\text{where } N_s = (1000 - \sum C_i/\varrho)/MW_s \tag{6}$$

$$\text{and } X_s = N_s/[N_s + \sum (\beta_i \times C_i/\varrho \times MW_i)] \tag{7}$$

The exergy change across the MD module or any other component of the MD system is calculated from the differences in inlet and outlet exergies as

$$\Delta Ex = \sum Ex_{in} - \sum Ex_{out} \tag{8}$$

The second-law efficiency of the MD system is the ratio of the minimum exergy input required (which is equivalent to the minimum work of separation) to the total actual exergy input:

$$\eta(\%) = (W_{min}/E_{xin}) \times 100 \tag{9}$$

The minimum work input required for the MD process is the difference between outgoing and incoming total exergies of the hot stream:

$$W_{min} = Ex_{permeate} + Ex_{feed\ out} - Ex_{feed\ in} \tag{10}$$

The exergy destruction due to irreversibility in the process is calculated by subtracting exergy determined from the second-law entropy generation [1] according to the Gouy–Stodola formula:

$$Ex_{dest} = S_{gen} \times T_o \tag{11}$$

where T_o is the reference temperature, and S_{gen} is the entropy generated due to irreversibility of the process, given as

$$S_{gen} = \Delta S_{total} = \Delta S_{source} + \Delta S_{sink} = Q_{source}/T_{source} + Q_{sink}/T_{sink} \tag{12}$$

where the Q_{source} and Q_{sink} represent the heat lost and gained from the heat source and sink, respectively.

The electrical exergy input for the pumps is

$$Ex_{el} = \dot{m} \times \Delta P/(1000 \times \eta_p) \tag{13}$$

where ΔP is the pressure head in Pa and η_p is for pump efficiency. The exergy of electric power involved for the pumps is equal to the energy and taken as pure exergy.

For simplicity, the exergy destroyed can be expressed as a percentage as related to the exergy input:

$$\varepsilon\% = Ex_{dest}/Ex_{in} \qquad (14)$$

$Ex_{dest}\%$ of a component is then calculated the fraction of exergy destruction by the particular component in comparison to the total destroyed exergy,

$$Ex_{dest}\% = \frac{Ex_{dest\ i}}{Ex_{dest\ total}} \qquad (15)$$

where $Ex_{dest\ i}$ is the exergy destroyed in component i, and $Ex_{dest\ total}$ is the total exergy destroyed in the MD system.

3. Results and Discussion

Exergy Efficiency of the AGMD Modules

Results from the permeate production rates for the two types of AGMD units are summarized in Figure 5 in terms of kilogram per square meter of membrane per second (kg/m^2s). The flux decreases as the temperature difference decreases across the MD module, which is due to the decline in the rate of condensation and concomitant drop in transmembrane transport of vapor across the membrane as the temperature difference decreases. Higher specific heat demands accompany this increase in permeate flux with increasing ΔT. The lower feed temperature condition generally shows a lower permeate flux, as observed for Xzero module's performance at 65 °C and 80 °C. When fluxes from the two MD modules are compared, the flux from Elixir500 MD unit is nearly three times higher than that from the Xzero MD unit at the same feed temperature and ΔT (65 °C).

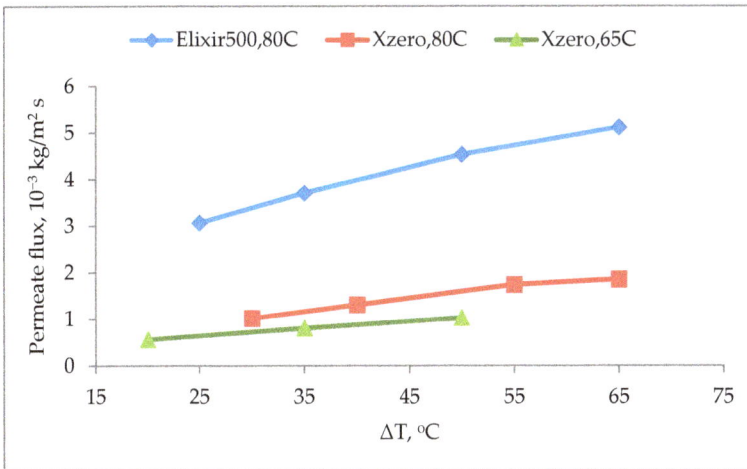

Figure 5. Permeate flux of Xzero and Elixir500 AGMD systems regarding permeate flux for different temperature differences at feed temperatures of 80 °C and 65 °C.

The other thermodynamic performance parameters: exergy efficiency, exergy destruction, and specific heat input are summarized for both modules in Figures 6 and 7. For both modules types, exergy efficiency improves when the feed–coolant temperature difference goes up. For the Xzero unit at a feed of 80 °C, exergy efficiency more than doubled as ΔT increased from 30 °C to 65 °C. This increase is linked to the improved heat recovery at the heat sink for a higher ΔT. This is evidently observed from the decrease in exergy destruction at a higher ΔT compared to lower levels (Figure 7). The trend is similar for both feed temperatures of 80 °C and 65 °C. However, the lower feed temperature condition generally shows lower exergy efficiency and higher exergy destruction when compared to the corresponding performance at higher feed temperatures. For this MD unit, it can be said that a higher feed temperature and ΔT favor higher flux and exergy efficiency and lower exergy destruction than lower ΔT cases, of course at the expense of additional specific heat demand.

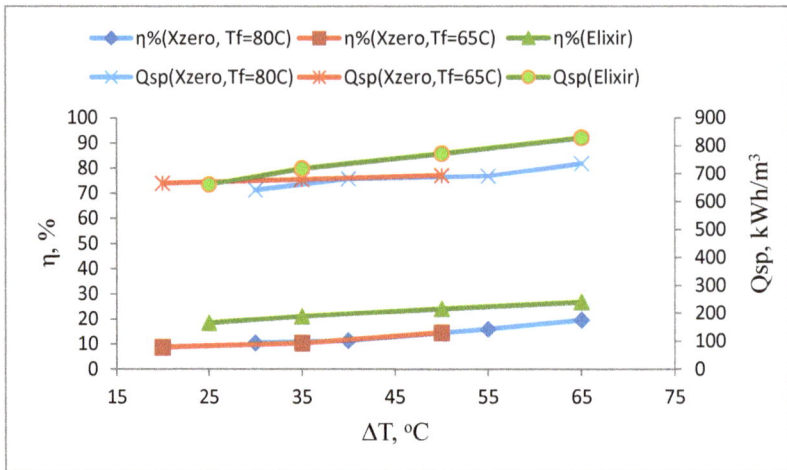

Figure 6. Performance of Xzero and Elixir500 AGMD systems regarding exergy efficiency and specific heat demand at different feed–coolant temperature differences for fixed feed temperatures of 80 °C (for both modules) and 65 °C (only for Xzero).

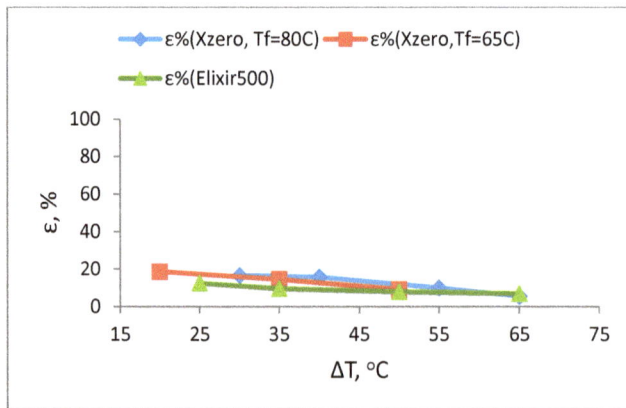

Figure 7. Exergy destructions and specific heat demand at different feed–coolant temperature differences for fixed feed temperatures of 80 °C (for both modules) and 65 °C (only for Xzero).

Similarly, for the bench scale AGMD unit, the general trend in the performances with a change in ΔT is the same as that of the results from the pilot-scale AGMD system. The effect of improved condensation in the Elixir500 module is reflected in the higher specific heat input owing to the higher heat transfer across the module. Hence, higher exergy efficiency and lower exergy destruction is achieved for the Elixir500 unit as compared to the Xzero module. The lower specific heat demand for the Xzero module results from the reduced heat transfer across the condensation plate in comparison to the other module, though the Xzero showed severe heat losses from the frames and cover to the surrounding atmosphere by free convection [8]. For both types of MD systems, it is evident that a lower temperature difference across the module reduces performance, including the permeate flux and exergy efficiency.

As the pressure drop across the AGMD modules is low (0.01 to 0.02 MPa) and mass balance is achieved across the MD unit, exergy losses from pressure and concentration changes are not significant. Hence, the major contributor to the exergy destruction is the heat loss from the modules to the surroundings and the latent heat transferred with the permeate. The heat transfer rates across the module can be assisted by employing condensation plates with good heat conduction, minimizing heat losses to the module frames and covers, and using membranes of lower heat conduction properties.

Considering each component of the MD system as open, the exergy destruction contributions can be pinpointed. Doing so will enable system designers to optimize the performance of the components. The result from calculations of exergetic destruction contributions of each component from both MD systems is summarized in Figure 8. The maximum fraction of exergy destruction in both MD types comes from the MD modules. However, the percentage is higher for the Xzero module (58%) than for the Elixir500 (43%). The pumps are in both cases the lowest contributors of exergy destruction. The re-concentrating tank contributed 34% of the exergy destruction to the Elixir500 unit, which was caused by the heat losses through the steel wall and evaporation occurred through openings in the cover as source of heat was electric heaters. Even though in both types of MD modules it is found that the MD modules contribute the highest exergy destruction, the Elixir500 modules showed a lower percentage than those of the Xzero. The lower percentage of exergy destruction for the Elixir500 MD module is mainly due to the improved heat transfer capability of the condensation plates used and hence the higher degree of heat recovery from the cooling side of the module.

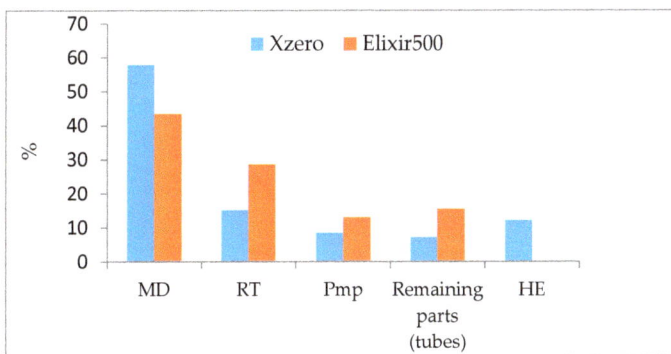

Figure 8. Percentage contribution of exergy destruction from the different components of each AGMD system.

A literature search for exergy efficiency analyses for related water purification processes, such as RO, revealed wide discrepancies in the values. Table 2 summarizes exergy efficiency for some of the standalone and integrated processes. In contrast to MD, the majority of the exergy rates in RO are derived from the high pressures involved, i.e., 15–25 bars for brackish water and 60–80 bars for seawater, hence the related pressure drop as reflected in improved efficiencies when a pressure exchanger is considered [19]. The very low exergy efficiency of MD reported [4] was from the one powered by solar, and the high exergy losses were due to the MD and heat exchanger units. The MD units' exergy efficiency values are relatively closer to each other than with RO. If we make comparisons between systems of similar processes and capacities, we see, for example, that the MD with a capacity of 0.3 m^3/day [4] and Elixir500 from the present study have comparable exergy efficiency.

Table 2. Summary of exergy efficiencies for different MD configurations and RO from the literature.

Process	Capacity, m^3/Day	Exergy Efficiency, %	Reference
RO	7250	4.3	[20]
RO	2850	0.72	[21]
SWRO	7586	5.82	[22]
MF-NF-RO	12,408	30.9	[23]
MD on RO retentate	22,344	19.1–21.9	[23]
MD	0.31	0.3	[4]
DCMD with HR	24,000	28.3	[11]
DCMD without HR	24,000	25.6	[11]
AGMD (Xzero)	0.22–0.73	8.54–19.32	(This study)
AGMD (Elixir500)	0.1–0.17	18.3–26.5	(This study)

4. Concluding Remarks

The exergy efficiency analyses that were carried out on pilot scale and bench MD systems illustrate the performance of the separation technologies based on useful energy, i.e., exergy. The AGMD systems tested showed different exergy efficiencies mainly due to the differences in module size and the type of condensation plates used. The exergy efficiencies were also affected to different degrees by the feed–coolant temperature differences in both types of modules. For the Xzero MD module, the exergy efficiency was not significantly affected by the difference in temperature across the module, whereas, for Elixir, a higher temperature difference across the module yielded higher exergy efficiencies and lower exergy destruction. The exergy efficiency results showed that the materials selection of the condensation plate plays a role in optimizing the performance of MD systems. This is clearly an important part of the MD system design and optimization, as the MD module contributes to the majority of the exergy destruction. Materials that could be considered for optimum heat transfer across modules and hence less exergy destruction from MD modules include stainless steel and high density polyethylene, which have better thermal conductivities than polypropylene. The selection of membrane and support material should also be considered when designing MD modules, as membrane and support materials, unlike condensation plates, should have less thermal conductivity for better thermal efficiency of modules.

Acknowledgments: This research has been done in collaboration with KTH Royal Institute of Technology, and Politecnico di Torino, PoliTo funded through Erasmus Mundus Joint Doctoral Program SELECT+, the support of which is gratefully acknowledged.

Author Contributions: Daniel Woldemariam conceived, designed, and performed the experiments; all three authors analyzed the data; and Daniel Woldemariam composed the article under the guidance of the co-authors.

Conflicts of Interest: The authors declare no conflict of interest. The funding sponsors had no role in the design of the study; in the collection, analyses, or interpretation of data; in the writing of the manuscript; or in the decision to publish the results.

References

1. Tsatsaronis, G. Definitions and nomenclature in exergy analysis and exergoeconomics. *Energy* **2007**, *32*, 249–253. [CrossRef]
2. Tsatsaronis, G. Thermoeconomic analysis and optimization of energy systems. *Prog. Energy Combust. Sci.* **1993**, *19*, 227–257. [CrossRef]
3. Guillén-Burrieza, E.; Blanco, J.; Zaragoza, G.; Alarcón, D.-C.; Palenzuela, P.; Ibarra, M.; Gernjak, W. Experimental analysis of an air gap membrane distillation solar desalination pilot system. *J. Membr. Sci.* **2011**, *379*, 386–396. [CrossRef]

4. Banat, F.; Jwaied, N. Exergy analysis of desalination by solar-powered membrane distillation units. *Desalination* **2008**, *230*, 27–40. [CrossRef]
5. Koschikowski, J.; Wieghaus, M.; Rommel, M.; Ortin, V.S.; Suarez, B.P.; Betancort Rodríguez, J.R. Experimental investigations on solar driven stand-alone membrane distillation systems for remote areas. *Desalination* **2009**, *248*, 125–131. [CrossRef]
6. Koschikowski, J.; Wieghaus, M.; Rommel, M. Solar thermal driven desalination plants based on membrane distillation. *Desalination* **2003**, *156*, 295–304. [CrossRef]
7. Khayet, M. Solar desalination by membrane distillation: Dispersion in energy consumption analysis and water production costs (a review). *Desalination* **2013**, *308*, 89–101. [CrossRef]
8. Woldemariam, D.; Kullab, A.; Fortkamp, U.; Magner, J.; Royen, H.; Martin, A. Membrane distillation pilot plant trials with pharmaceutical residues and energy demand analysis. *Chem. Eng. J.* **2016**, *306*, 471–483. [CrossRef]
9. Dow, N.; Gray, S.; Li, J.; Zhang, J.; Ostarcevic, E.; Liubinas, A.; Atherton, P.; Roeszler, G.; Gibbs, A.; Duke, M. Pilot trial of membrane distillation driven by low grade waste heat: Membrane fouling and energy assessment. *Desalination* **2016**, *391*, 30–42. [CrossRef]
10. Jansen, A.E.; Assink, J.W.; Hanemaaijer, J.H.; van Medevoort, J.; van Sonsbeek, E. Development and pilot testing of full-scale membrane distillation modules for deployment of waste heat. *Desalination* **2013**, *323*, 55–65. [CrossRef]
11. Al-Obaidani, S.; Curcio, E.; Macedonio, F.; Di Profio, G.; Al-Hinai, H.; Drioli, E. Potential of membrane distillation in seawater desalination: Thermal efficiency, sensitivity study and cost estimation. *J. Memb. Sci.* **2008**, *323*, 85–98. [CrossRef]
12. Criscuoli, A.; Drioli, E. Energetic and exergetic analysis of an integrated membrane desalination system. *Desalination* **1999**, *124*, 243–249. [CrossRef]
13. Macedonio, F.; Drioli, E. An exergetic analysis of a membrane desalination system. *Desalination* **2010**, *261*, 293–299. [CrossRef]
14. Sow, O.; Siroux, M.; Desmet, B. Energetic and exergetic analysis of a triple-effect distiller driven by solar energy. *Desalination* **2005**, *174*, 277–286. [CrossRef]
15. García-Rodríguez, L.; Gómez-Camacho, C. Exergy analysis of the SOL-14 plant (Plataforma Solar de Almería, Spain). *Desalination* **2001**, *137*, 251–258. [CrossRef]
16. Banat, F.; Jwaied, N.; Rommel, M.; Koschikowski, J.; Wieghaus, M. Desalination by a "compact SMADES" autonomous solarpowered membrane distillation unit. *Desalination* **2007**, *217*, 29–37. [CrossRef]
17. Aljundi, I.H. Second-law analysis of a reverse osmosis plant in Jordan. *Desalination* **2009**, *239*, 207–215.
18. Khan, E.U.; Martin, A.R. Water purification of arsenic-contaminated drinking water via air gap membrane distillation (AGMD). *Period. Polytech. Mech. Eng.* **2014**, *58*, 47–53. [CrossRef]
19. Querol, E.; Gonzalez-Regueral, B.; Perez-Benedito, J.L. Exergy Concept and Determination. In *Practical Approach to Exergy and Thermoeconomic Analyses of Industrial Processes*; Springer: Heidelberg/Berlin, Germany, 2013; pp. 9–28.
20. Cerci, Y. Exergy analysis of a reverse osmosis desalination plant in California. *Desalination* **2002**, *142*, 257–266. [CrossRef]
21. Ameri, M.; Eshaghi, M.S. A novel configuration of reverse osmosis, humidification–dehumidification and flat plate collector: Modeling and exergy analysis. *Appl. Therm. Eng.* **2016**, *103*, 855–873. [CrossRef]
22. El-Emam, R.S.; Dincer, I.; Salah El-Emam, R.; Dincer, I.; El-Emam, R.S.; Dincer, I. Thermodynamic and thermoeconomic analyses of seawater reverse osmosis desalination plant with energy recovery. *Energy* **2014**, *64*, 154–163. [CrossRef]
23. Drioli, E.; Curcio, E.; Di Profio, G.; Macedonio, F.; Criscuoli, A. Integrating Membrane Contactors Technology and Pressure-Driven Membrane Operations for Seawater Desalination. *Chem. Eng. Res. Des.* **2006**, *84*, 209–220. [CrossRef]

applied
sciences

MDPI

Review

Fouling in Membrane Distillation, Osmotic Distillation and Osmotic Membrane Distillation

Mourad Laqbaqbi [1,2], Julio Antonio Sanmartino [1], Mohamed Khayet [1,3,*], Carmen García-Payo [1] and Mehdi Chaouch [2]

[1] Department of Applied Physics I, Faculty of Physics, University Complutense of Madrid,
    Avda. Complutense s/n, 28040 Madrid, Spain; mouradlaqbaqbi@gmail.com (M.L.);
    julio.sanmartino@hotmail.com (J.A.S.); mcgpayo@fis.ucm.es (C.G.-P.)
[2] Laboratory of Materials Engineering and Environment, Department of Chemistry, Faculty of Sciences Dhar
    El Mehraz, 30000 Fez, Morocco; mechaouch@yahoo.fr
[3] Madrid Institute for Advanced Studies of Water (IMDEA Water Institute), Avda. Punto Com 2,
    Alcalá de Henares, 28805 Madrid, Spain
* Correspondence: khayetm@fis.ucm.es; Tel.: +34-91-3964-51-85

Academic Editor: Bart Van der Bruggen
Received: 1 February 2017; Accepted: 24 March 2017; Published: 29 March 2017

Abstract: Various membrane separation processes are being used for seawater desalination and treatment of wastewaters in order to deal with the worldwide water shortage problem. Different types of membranes of distinct morphologies, structures and physico-chemical characteristics are employed. Among the considered membrane technologies, membrane distillation (MD), osmotic distillation (OD) and osmotic membrane distillation (OMD) use porous and hydrophobic membranes for production of distilled water and/or concentration of wastewaters for recovery and recycling of valuable compounds. However, the efficiency of these technologies is hampered by fouling phenomena. This refers to the accumulation of organic/inorganic deposits including biological matter on the membrane surface and/or in the membrane pores. Fouling in MD, OD and OMD differs from that observed in electric and pressure-driven membrane processes such electrodialysis (ED), membrane capacitive deionization (MCD), reverse osmosis (RO), nanofiltration (NF), ultrafiltration (UF), microfiltration (MF), etc. Other than pore blockage, fouling in MD, OD and OMD increases the risk of membrane pores wetting and reduces therefore the quantity and quality of the produced water or the concentration efficiency of the process. This review deals with the observed fouling phenomena in MD, OD and OMD. It highlights different detected fouling types (organic fouling, inorganic fouling and biofouling), fouling characterization techniques as well as various methods of fouling reduction including pretreatment, membrane modification, membrane cleaning and antiscalants application.

Keywords: membrane distillation; osmotic distillation; osmotic membrane distillation; fouling; organic fouling; scaling; biofouling; fouling characterization; fouling reduction; antiscalant

1. Introduction

The lack of potable water is one of the continuous problems in many parts of the word. Seawater desalination using isothermal membrane separation processes such as reverse osmosis (RO) and nanofiltration (NF) are a convincing solution. Pressure-driven membrane processes are limited in recovery factor due to the osmotic pressure, which increases with salinity, enhancing therefore water cost and environmental perturbations when the brines are not recycled and discharged directly in the feed water source (e.g., seas and rivers).

Membrane distillation (MD), osmotic distillation (OD) and osmotic membrane distillation (OMD) processes are used not only in desalination for water production and concentration of brines but also for the treatment of wastewaters (e.g., textile, radioactive, pharmaceutical, metallurgical, petrochemical, etc.) and concentration of heat-sensitive solutions such as fruit juices, liquid foods, natural colors and biological fluids since these processes operate at moderate temperatures and under atmospheric pressure.

MD is a thermally-driven separation process, in which only vapor molecules are transported through a microporous hydrophobic membrane. The MD driving force is the transmembrane vapor pressure difference [1–3]. Various MD configurations have been considered to apply this driving force. In direct contact membrane distillation (DCMD) both the hot feed solution and the cold liquid permeate are maintained in direct contact with both sides of the membrane. This is the widely employed MD variant because of its simple design [1]. In this case volatile molecules evaporate at the hot liquid/vapor interface, cross the membrane pores in vapor phase and condense in the cold liquid/vapor interface inside the membrane module [2]. Liquid gap MD (LGMD) is another DCMD variant in which a stagnant cold liquid, frequently distilled water, is maintained in the permeate side between the membrane and a cold surface [2]. The main disadvantage of DCMD and LGMD is the heat lost by conduction through the membrane [1]. If in LGMD the liquid is evacuated from the permeate side leaving a stagnant air gap between the membrane and the cold surface for the condensation of the volatile molecules, the configuration is termed air gap MD (AGMD) [2]. One of the advantages of this MD variant is the low heat transfer by conduction through the membrane from the feed to the permeate side [1]. However, in this case an additional resistance to mass transfer is built reducing the permeate flux. If in AGMD, a cold inert gas is circulated through the permeate side to carry out the produced vapor at the permeate membrane surface for condensation outside the membrane module in an external condenser(s), the configuration is called thermostatic sweeping gas MD (TSGMD) [2]. If the condensing surface is removed from the permeate side, the process is termed simply sweeping gas MD (SGMD). Therefore, TSGMD is a combination of AGMD and SGMD and it was proposed to reduce the increase of the gas temperature along the membrane module length [1]. Another way to establish the necessary driving force in MD is by means of a vacuum pump connected to the permeate side of the membrane module. This configuration is called vacuum MD (VMD) [2]. In this case the applied vacuum pressure must be lower than the saturation pressure of the volatile molecules to be separated from the feed solution and the condensation takes place outside the membrane module. It is necessary to point out that MD was applied principally in desalination for the production of high purity water. Other fields of applications have also been considered at laboratory scale such as the treatment of textile wastewater, olive mill wastewater, humic acid (HA) and radioactive aqueous solutions, etc. [2].

Contrary to DCMD, osmotic distillation (OD) is an isothermal technology used to remove water from aqueous solutions [4] (i.e., concentration of wastewaters and recovery of valuable components) [5] using a porous and hydrophobic membrane that separates the feed solution to be treated and an osmotic solution (i.e., a draw solution having high osmotic pressure and low water chemical potential) [6]. Generally NaCl, CaCl$_2$, MgCl$_2$, MgSO$_4$, K$_2$HPO$_4$ KH$_2$PO$_4$ and some organic liquids like glycerol or polyglycols were considered to prepared the osmotic solution [7]. One of the advantages of the OD process is the less energy required compared to MD [8]. It was generally applied for concentrating liquid foods, such as milk, fruit and vegetable juices [5].

OMD is a combination of DCMD and OD being the driving force both the transmembrane temperature and concentration or which is the same water chemical potential. It is a non-isothermal process in which the membrane is brought into contact with the hot feed aqueous solution to be treated and a cold osmotic solution [9]. OMD is also able to concentrate liquid foods (i.e., fruit and vegetable juices), sucrose aqueous solutions [10].

The three separation processes MD, OD and OMD require the use of porous and hydrophobic membranes and, like other membrane technologies, they also suffer from different fouling phenomena that reduce not only the permeability and separation performance of the membrane but its lifetime as a consequence. The foulants (e.g., natural organic matter, NOM [11]; inorganic and biological solutes or microbial contaminants) contribute to the permeate flux decline, modify the membrane surface properties and change the product water quality. Analysis of fouling process and identification of foulants by means of characterization techniques are important to determine suitable treatment methods for fouling control. It is worth noting that very few studies have been published so far on fouling mechanisms in MD, OD and OMD and investigations on the kinetics behind fouling phenomena and fouling mitigation remain very scarce [12,13]. Two review papers have been published on fouling and scaling in MD but not on OD and OMD [14,15]. Moreover, when fouling is studied, the considered characterization techniques focused only on the average physico-chemical properties of the surface deposits but not on the underlying deposit layers [12].

One of the causes of porous and hydrophobic membrane fouling is pore blockage due to the deposit of foulant(s) such as organics, inorganic or minerals, colloids, microbial contaminants and particles not only on the membrane surface but also inside the membrane pores affecting the hydrophobic character of the membrane and its wettability [16,17]. Fouling in MD, OD and OMD is a complex phenomenon influenced by various parameters such as the membrane characteristics, especially the pore size and the material of the membrane surface, operation conditions and nature of feed aqueous solutions.

To control fouling phenomena, researchers have tried various strategies such as the consideration of feed pretreatment(s), increase of feed flow rate creating turbulent flow regime, application of periodic hydraulic and/or chemical cleanings, reduction of membrane surface roughness and/or change of its surface charge [16,18].

It must be mentioned that MD, OD and OMD membrane technologies suffer from temperature and/or concentration polarization, or, equivalently, vapor pressure polarization. Various strategies have been adopted in order to reduce the vapor pressure polarization (i.e., the water vapor pressure at the membrane surface become closer to that of the bulk solution) including the increase of the flow rate of both the feed and permeate solutions, turbulent promoters, etc. It must be noted that concentration and temperature polarization can also have a major influence on fouling.

This review deals with the observed fouling phenomena in MD, OD and OMD. It highlights different detected fouling types (organic fouling, inorganic fouling and biofouling), fouling characterization techniques as well as various methods of fouling reduction including pretreatment, membrane modification, membrane cleaning and antiscalants application. Updated research studies and interesting observations on fouling and pretreatments are summarized in tables for different MD configurations, OD and OMD. Not only the characterization techniques that have been used so far in MD are cited in the present review, but other useful techniques for fouling analysis and detection considered in other processes are included. These will improve the understanding of fouling in MD, OD and OMD, to prevent it properly.

2. Membrane Characteristics

The membranes used in MD, OD and OMD must be hydrophobic and porous with pore sizes ranging from some nanometers to few micrometers [2,19]. Their characteristics such as the thickness, tortuosity, pore size and porosity dictate the resistance to mass transfer in these three processes [20]. Their pore size distributions should be as narrow as possible and the maximum pore size should be small enough to prevent liquid penetration in such pores [2,19,21]. The liquid entry pressure (*LEP*), which is the minimum transmembrane pressure required for a liquid or a given feed solution to enter into the pore, is a significant membrane characteristic for MD, OD and OMD. *LEP* is high for small maximum pore sizes and more hydrophobic membranes [1]. The membrane thickness is inversely proportional to the rate of mass and heat transfer through the membrane. In the case of multi-layered membranes, the hydrophobic layer should be as thin as possible [2,19]. In order to achieve a high thermal efficiency in MD and OMD, the thermal conductivity of the membrane material should be as low as possible [19].

More details on the properties needed for a membrane to be used in MD are summarized elsewhere [15,18,19,22,23]. Eykens, et al. [23] gave a comprehensive overview of the optimal membrane properties, specifically for MD process. The wetting resistance is the key factor considered in the optimization study. The recommended optimal membrane properties were summarized as a pore diameter of 0.3 µm to balance between a high *LEP* (preferably >2.5 bar) and a high permeate flux, an optimal membrane thickness between 10 and 700 µm depending on the process conditions in order to take into account the compensation between mass transport and energy loss, a membrane porosity that should preferably be as high as possible (>75%) in order to improve both the mass transfer and energy efficiency, a pore tortuosity factor that should be as low as possible (1.1–1.2) and a membrane thermal conductivity that should be also as low as possible (>0.06 $W \cdot m^{-1} \cdot K^{-1}$) in order to reduce the heat loss due to the heat transfer by conduction through the membrane matrix. Additionally, it was stated that thinner membranes with a thickness below 60 µm exhibited low mechanical properties. As it is well known, a way to improve the mechanical properties of a membrane without scarifying its other characteristics is the design of supported membranes using baking materials with a high porosity, a low thickness and a high thermal conductivity.

In addition, the feed side of the membrane must be formed by a material of high fouling resistance properties. Different membrane surface modification techniques have been considered such as coating, interfacial polymerization, plasma treatment, etc. [2,19,24]. For instance, to avoid wetting and membrane fouling the porous hydrophobic membrane surface needs to be modified by coating a thin layer of a hydrophilic polymer [24–26]. This was necessary for concentration of oily feeds, because the uncoated membranes were promptly wetted even for low concentrations of oil in water solutions [25]. Recently, the effects of surface energy and its morphology on membrane surface omniphobicity (i.e., membranes resistant to wetting to both water and low surface tension liquids, e.g., oil and alcohols) have been systematically studied by means of wetting resistance evaluation using low surface tension liquids [27,28]. It was found that the negatively charged nanofibrous membrane fabricated by a blend of poly(vinylidene fluoride-*co*-hexafluoropropylene), PVDF-HFP, and the cationic surfactant benzyltriethylammonium, and then grafted by a negatively charged silica nanoparticles, exhibited excellent wetting resistance against low surface tension aqueous solutions and organic solvents (i.e., mineral oil, decane and ethanol). An et al. [29,30] demonstrated that the MD membranes having negative charge such as PTFE at pH values in the range 5.2–9.1 were resistant to dye adsorption. The strong negative charge and chemical structure of the membrane resulted in a low adsorption affinity to negatively charged dyes causing a flake-like (loose) dye-dye structure to form on the membrane surface rather than in the membrane pores or even repulsed from the membrane forming aggregates away from the membrane interface. In addition, the loose fouling structure that may be formed on the surface of the negatively charged membranes can be easily washed out by simple intermittent water flushing. It is to be noted that the membrane must be cleaned if any fouling is detected and therefore it should exhibit an excellent chemical resistance to acid and base components that are generally used in membrane cleaning [19].

3. Fouling in MD, OD and OMD

The performance of the membrane can be affected by the deposition of a fouling layer on the membrane surface or in the membrane pores [31]. The decline of water permeate flux is attributed to both temperature and concentration polarization effects as well as fouling phenomena [32].

3.1. Fouling in MD

Compared to fouling in other membrane separation processes such as the pressure-driven membrane processes (MF, NF, RO, etc.) fouling in MD is still relatively less studied and poorly understood [33–35]. In MD, fouling can be divided into three types: organic fouling, inorganic fouling and biological fouling. In general, the foulants interact with each other and/or with the membrane surface to form deposits. This results in permeate flux decline by two phenomena, a partial or total blockage of the pores, which decreases the available evaporation area; or the formation of a fouling layer on the membrane surface leading to the appearance of a new resistance to mass transfer. As a consequence, the membrane becomes more prone to wetting, especially for long term MD operations. It is worth noting that most of the published studies on fouling phenomena in MD are for DCMD configuration. Table 1 summarizes the published papers on fouling in DCMD when using different membrane materials (polytetrafluoroethylene, PTFE, polyvinylidene fluoride, PVDF, and polypropylene, PP, polymers) and membrane type (flat-sheet, hollow fiber or capillary membranes). Very few studies have been published on fouling in other MD configurations. Table 2 listed the published papers on fouling in AGMD and VMD configurations. In order to compare the effects of the foulants, membranes, MD configurations, etc. the normalized flux decline, FD_n, defined in the following equation was used:

$$FD_n(\%) = \left(1 - \frac{J_f}{J_0}\right) \times 100 \tag{1}$$

where J_0 and J_f correspond to the initial and final permeate fluxes, respectively. As it can be seen in Tables 1 and 2, the FD_n values varied between 0% and 95%. These values are strongly dependent on the foulant nature. In general, organic foulants induced greater normalized flux decline than the inorganic foulants.

Table 1. Published studies on membrane fouling in DCMD for different membrane materials and types (d_p: pore size; ε: porosity; δ: thickness; T_f: Feed temperature; T_p: Permeate temperature; v: feed velocity (or ϕ: feed flow rate); J_p: Permeate flux; FD_n: Normalized flux decline; F: Feed; P: Permeate).

Membrane Material and Type	d_p (μm)	ε (%)	δ (μm)	Foulant(s)	T_f/T_p (°C)	v (m·s⁻¹)	J_p (kg·m⁻²·h⁻¹)	FD_n (%)	Observations	Ref.
	0.1	-	30	Organic fouling: sludge and brown spots	40–70/10	0.005–0.014	1.41–9.22	70% at 12g/L 50% at 6 g/L	The feed was from a thermophilic anaerobic membrane bioreactor. After cleaning using deionized (DI) water 15 mg/L of NaOH, the membrane could recover 96% of initial J_p.	[36]
	0.5	-	20	Skim milk, Whey proteins	54/5	0.047	3	Skim milk: 85% Whey: 20%	Membrane wetting was not observed even after 20 h operation.	[37]
	0.2	80	60	Traditional Chinese medicine (mostly inorganic salt, such as CaCO₃)	60/25	0.07–0.13	32.78	30% at 1.5 g/L	Fouling layer can be effectively limited by increasing either the feed temperature or feed flow velocity.	[38]
PTFE Flat-sheet	0.2	70–80	179	Humid Acid (HA)	70/25	1.1	35.7	60%	Seawater organic fouling was irreversible in DCMD. CaSO₄ reduced the disaggregation of humic substances due to the binding effect.	[39]
	0.5	-	20	Skim milk Whey proteins	54/5	0.047	22	Skim milk: 79% Whey: 11%	Whey proteins had weaker attractive interaction with the membrane and adhesion depended more on the presence of phosphorus near the membrane surface.	[12]
	0.2	70–80	179	HA, alginate acid (AA) and bovine serum albumin (BSA)	50, 70/24	1.1	35	HA: 56.2% AA: 44.1% BSA 64.5%	Feed concentration: 10 mg·C·L⁻¹.	[40]
	0.2	65	41				~28, at ΔT = 15 °C	28%	The PTFE membrane surfaces showed some salt scaling and a larger population of crystals.	
	0.2	80	197	NaCl, MgSiO₃, MgCO₃ CaCO₃	30–50/24	0.32	~5, at ΔT = 15 °C	32%	PVDF membranes showed a smaller population of salt crystals on their surface.	[41]
PVDF Flat-sheet	0.45	60	127				~32, at ΔT = 15 °C	20%	The salt crystals formed were larger than the pore size of the membranes.	
	0.22	0.75	125	HA	50, 70/20	0.23	30.6 to 35.1	5%	The addition of divalent cations (Ca²⁺) affected to permeate flux by forming complexes with HA.	[42]
	0.16	90.8	200	CaSO₄ (RO brine)	55–77/35	0.011	2.5–5.8	30%	Membrane wetting was more significant at high feed temperatures. Salts promoted membrane pore wetting.	[43]
PVDF Hollow fiber	0.088	83.7	126	Rubber wastewater	55.5/20.0	-	7.19	79%	Permeate flux decline was due to the presence of complex components (e.g., latex and protein in the rubber effluent).	[44]

Table 1. *Cont.*

Membrane Material and Type	d_p (μm)	ε (%)	δ (μm)	Foulant(s)	T_f/T_p (°C)	v (m·s⁻¹)	J_p (kg·m⁻²·h⁻¹)	FD_n (%)	Observations	Ref.
PP Flat-sheet	0.1	65–70	100	CaCO₃, CaSO₄, silica	60/20	φ: 0.6/0.5 L·min⁻¹	30	Feed A: 93% Feed B: 27%	Membrane scaling caused a drop of permeate flux and a decrease in salt rejection. Feed A: NaHCO₃ + Na₂SO₄ + CaCl₂; Feed B: Na₂SiO₃·9H₂O.	[45]
PP Hollow fiber	0.22	73	400	CaCO₃	85, 90/20	F: 0.15–0.63 P: 0.12	25–38.7	40% without antiscalant 20% with antiscalant 0% Rising with HCl solution	The application of antiscalant minimized the penetration of salts into the pores. A high permeate flux was maintained over 260 h of operation using periodical rinsing with HCl solution	[46]
	0.1 0.2 0.6	50 60–80	150 50 52.5	CaCO₃ and CaCO₃-CaSO₄	70-75-80/20	φ: 0.084-0.688-1.438 L·min⁻¹	Uncoated: 14.3-4.8 Coated: 5.5	11%	Fluorosilicone coating was proven to be helpful to eliminate membrane scaling. The scaling problem was successfully solved by HCl acidification prior to MD	[35]
	0.22	73	400	CaCO₃	60-80/50	F: 0.42–0.96 P: 0.29	27.9–22.9	10%	Reduction of the number and dimensions of the pores on the membranes surfaces. Large pores were wetted because of CaCO₃ deposition inside the pores. Flow rate of distillate was constant.	[47]
PP Capillary	0.22	73	400	Protein and NaCl (50 g/L)	85/20	φ: 0.84 L·min⁻¹	12.5	25% NaCl up to saturation 0% Boiling feed	The feed solution was NaCl solution containing natural organic matter. Pre-treatment method of the feed does not result in the complete removal of the foulants.	[48]
	0.22	73	400	Mainly CaCO₃	80, 90/20	F: 0.3–1.4 P: 0.29	30.8 (T_f = 80 °C)	41% at 0.31 m/s 12% at 1.4 m/s	The application of tap water as a feed caused a rapid decline of permeate flux due to the deposition of CaCO₃.	[49]
	0.22	72	200	NaCl	70-85/20	φ: 0.42 L·min⁻¹	27.5	19% at 1 year 28% at 9 years	Chemical reaction of salt with the hydroxyl and carbonyl groups found on the PP surface. Degradation time dependence.	[50]

Appl. Sci. 2017, 7, 334

Table 2. Published studies on membrane fouling in AGMDVMD for different membrane materials and types (d_p: pore size; ε: porosity; δ: thickness; T_f: Feed temperature; T_p: Permeate temperature; v: feed velocity (or ϕ: feed flow rate); I_p: Permeate flux; FD_n: Normalized flux decline; F: Feed; C: Cooling; p_0: Vacuum or low pressure) [51–59].

MD Config.	Membrane Material and Type	d_p (µm)	ε (%)	δ (µm)	Foulant	T_f/T_p (°C) p_0 (kPa)	v (m·s⁻¹)	I_p (kg·m⁻²·h⁻¹)	FD_n (%)	Observations	Ref.
AGMD		1	0.85	150	Salts deposition	77/12	ϕ: F: 58 g/s C: 75 g/s	119	-	Feed: tap water.	[51]
	PTFE Flat-sheet	0.2	-	175	NaCl, MgCl₂, Na₂CO₃, Na₂SO₄	50/10	ϕ: F: 1.5 L·min⁻¹	NaCl: 1.02 Na₂SO₄: 0.38 Na₂CO₃: 0.12	0% after 5 h	The permeate flux declined as the concentration of salt increased, and increased as the pore size increased.	[1]
		0.45	-	175	NaCl, MgCl₂, Na₂CO₃, Na₂SO₄	50/10	ϕ: F: 1.5 L·min⁻¹	NaCl: 1.45 Na₂SO₄: 0.56 Na₂CO₃: 0.19	0% after 5 h		
		0.22	40	175	CaCO₃, CaSO₄ organic matter	25–75 p_0: 0.1–10	0.4–2.0	9.3–8.3	24%	Synthetic RO brine feed, Salt concentration: 300 g·L⁻¹.	[53]
		0.2	-	50	Ginseng aqueous solution (polysaccharide, amino acid & biomacromolecule)	60 p_0: 83–89.5	0.74 to 0.46	24.7 to 21.6	From 0% to 27%	The results showed the existence of critical fouling operating conditions in VMD process.	[54]
VMD	PVDF Hollow fiber	0.25	79	150	CaCO₃	52–68 p_0: 96	0.14	8.96–21 (13.43–25 with microwave)	-	Microwave irradiation increased the deposition of calcium carbonate.	[55]
		0.16	82–85	-	Mainly hardness and organic matter	70 p_0: 8.5	1	-	-	The permeate flux was 30% higher when using pretreatment.	[56]
	PP Shell and tube	0.2	-	1500	Dye and Nacl	40–70 p_0: 0.667	0.84 to 3.42	8.2	95%	The flux was dependent strongly on the feed temperature but was independent of salt concentration.	[57]
	PP Flat-sheet	0.2	-	-	Inorganic elements: O, S, Fe, Na, Mg, K. Microorganisms and proteins,	40 °C	ϕ: F: 4 L·min⁻¹ Strip flow rate: 3 L·min⁻¹	42	95%	The fouling layer thickness was estimated to 10–15 µm and it becomes severe as the membrane surface changes from hydrophobic to hydrophilic.	[58]
	PP Capillary	0.2	45	510	Dye	40, 50, 60 p_0: 1.0	0.78–1.67	27.5–35–57.0, respectively	27%	The flux decrease probably due to an interaction with the polymeric membrane.	[59]

3.1.1. Organic Fouling

Organic fouling results from the deposition of natural organic matter (NOM) on the membrane surface like carboxylic acid, humic acid (HA), alginate acid (AA), proteins, polysaccharides, etc. The principal constituents of NOM are the humic substances, which are found in surface water, ground water and seawater, followed by carbohydrates (including polysaccharides), protein and a variety of acidic and low molecular weight (LMW) species [40,60,61]. This fouling is dependent on several factors including the surface characteristics of the membrane. For instance, greater hydrophobicity and lower pore size tends to increase fouling effects [31]. However, fouling also depends on the nature of the organic matter, the MD operating conditions (temperature, transmembrane pressure, flow rate) and the characteristics of the feed solution (pH, ionic strength). Figure 1 shows the adsorption–desorption mechanism for HA migration through a membrane pore. The process involves the adsorption of HA onto membrane surface, hydrogen bonding between water and HA and weakening of hydrogen bond as water vapor moves through the membrane, and re-adsorption of HA onto the membrane as well as pore wetting.

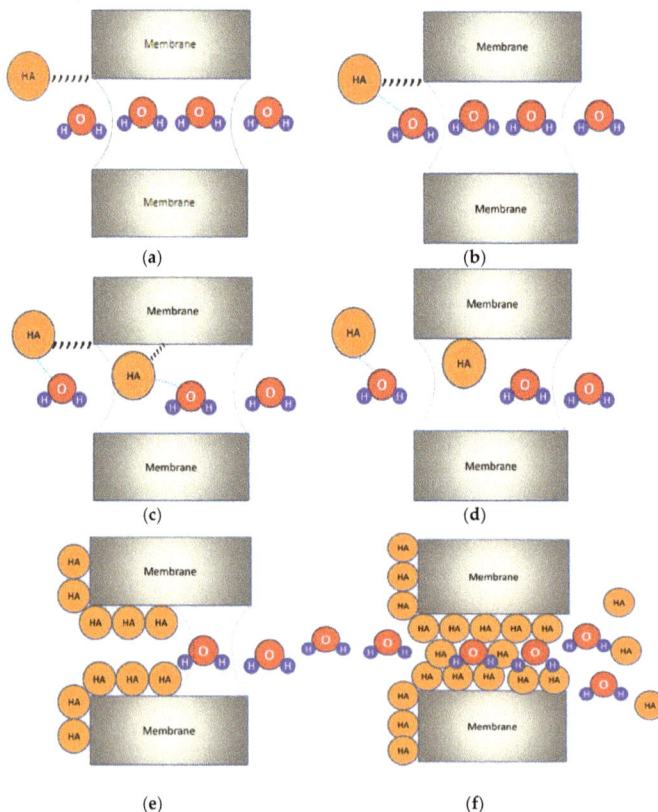

Figure 1. Illustration of the adsorption–desorption mechanism for humic acid (HA) migration through a membrane pore. The process involves (**a**) adsorption of HA onto membrane surface; (**b**) hydrogen bonding between water and HA; (**c,d**) weakening of hydrogen bond as water vapor moves through the membrane; (**e**) re- adsorption of HA onto the membrane and (**f**) the pore wetting phenomenon.

Nilson and Digiano [62] investigated the effect of NOM properties on NF membrane fouling by fractionating NOM into hydrophilic and hydrophobic components. Fouling tests revealed that the hydrophobic fraction of NOM was mostly responsible for permeate flux decline, whereas the hydrophilic fraction caused a lesser fouling. Khayet, et al. [11,31] performed DCMD treatment by HA solutions (10 to 50 mg·L^{-1}) and the obtained results showed a permeate flux reduction less than 8% after 30 h of operation for the commercial PTFE and PVDF membranes. Other studies used HA solutions with concentrations ranging from 20 to 100 mg·L^{-1} and observed a very limited fouling with a permeate flux decline less than 6% [42]. The reversibility or cleaning of organic fouling was also investigated but the drawn conclusions varied because of the possible variety of organic fouling such as organic or organometallic complex (i.e., compound containing a metal atom bonded to an organic group). Naidu, et al. [40] studied organic fouling behavior in DCMD using synthetic model solutions of HA, AA and bovine serum albumin (BSA). BSA and HA showed a dominant fouling with a permeate flux decline of 50%, whereas AA showed minimal fouling due to its hydrophilic nature [40]. When comparing BSA and HA feed solutions, higher deposits were detected for BSA feed solution (i.e., 35.2% higher carbon mass); and the analysis of fouled membrane proved the penetration of organic compounds through the membrane pores when using HA feed solution [40].

It was demonstrated that the aggregation of the HA and the increase of membrane fouling is favored with the addition of multivalent cations and the increase of electrolyte (NaCl) concentration in the feed aqueous solution. Multivalent cations including Calcium (Ca^{2+}) are known to form complexes with NOM and interact specifically with humic carboxyl functional groups reducing the humic charge and the electrostatic repulsion forces between humic macromolecules resulting in aggregates in NF or MF processes [60,63].

Some studies showed the irreversibility of organic fouling in DCMD [11,31]. However, other studies claimed the reversibility/cleaning of organic fouling, e.g., the complexes formed by calcium ion (Ca^{2+}) and organic matter that precipitates only on membrane surface forming a thin deposit layer, which is eliminated completely by a simple cleaning with water and with 0.1 M sodium hydroxide solution [42]. It was reported that cleaning the membrane by a simple flushing with deionized water through the feed channel at a flow velocity of around 1 m·s^{-1} permitted the recovery of up to 98% of permeate flux [40].

Gryta [64] proved that organic fouling can be prevented if a specific pretreatment is applied. Before the MD treatment of wastewater originated from heparin production from intestinal mucous, Gryta [64] separated these materials by boiling during 30 min followed by their separation after filtration. Srisurichan, et al. [42] studied the HA fouling in MD and reported that greater fouling and severe permeate flux decline (i.e., ratio of final and initial permeate fluxes, $J/J_0 = 0.57$) was observed after 18 h of operation in presence of high concentration of CaCl$_2$ ($\approx$3.775 mM) in which large amounts of HA were present in coagulate form.

3.1.2. Inorganic Fouling

Inorganic fouling in MD is due to the precipitation and crystallization of salts present in the used aqueous feed solutions. In the case of desalination of seawater, crystallization fouling is attributed mainly to sodium chloride (NaCl), which is the predominant salt in the feed solution other than the divalent ions such as calcium or magnesium salts. It is worth noting that sodium chloride can precipitate as halite (i.e., cubic crystals of NaCl), while calcium sulfate precipitates following its hydration, in the form of anhydrite (CaSO$_4$), hemihydrate (CaSO$_4 \cdot \frac{1}{2}$H$_2$O) or gypsum (CaSO$_4 \cdot 2$H$_2$O) [65]. Calcium carbonate (CaCO$_3$) precipitates as calcite (cubic crystals of CaCO$_3$) and it is usually found in the form of rhombohedron, which is the most thermodynamically stable form of all other varieties such as aragonite (less stable than calcite) and vaterite (spherical crystals of CaCO$_3$) [65].

It is to be noted that the concentration and temperature polarization may have a major influence on inorganic fouling in MD. For a salt whose solubility decreases with increasing temperature (e.g., CaCO$_3$, CaSO$_4$, BaSO$_4$) the temperature polarization phenomenon induces the formation of these salt(s) crystals in the bulk feed and not on the membrane surface provided that the temperature at the membrane surface is lower. On the other hand, for salts whose solubility increases with increasing temperature (e.g., NaCl), the concentration and temperature polarization phenomena encourage the formation of crystals on the membrane surface, where the concentration is higher and temperature is lower and not in the bulk feed, where the temperature is higher and concentration is lower.

Basically, the carbonic system is derived from the dissolution of carbon dioxide and carbonate minerals in water. Thus, carbonate is a weak acid-base system, which exists in aqueous solutions as dissolved carbon dioxide CO_{2aq}. carbonic acid (H_2CO_3), bicarbonate (HCO_3^-), carbonate ions (CO_3^{2-}) and complexes of these ions such as $CaHCO_3^+$,$CaCO_3$ [65]. Dissolved carbon dioxide is hydrated in a few minutes according to the reaction ($CaCO_{3aq} + H_2O \Leftrightarrow CO_2, H_2O$). The term H_2CO_3 refers to the composite form, which is the sum of the activities of molecularly dissolved carbon dioxide CO_{2aq}, and the hydrated form CO_2, H_2O.

In MD literature, research studies on inorganic fouling have focused on the treatment of aqueous solutions containing NaCl by VMD using hollow fiber membranes [66] or by DCMD using flat sheet membranes [67]. It was detected a limited permeate flux decline of about 30% to 35% at concentrations ranging from 15 to 300 g/L. When the salts crystallize on the membrane surface, these can cover the pores and reduce the effective area available for vapor-liquid interface reducing subsequently the permeate flux. On the other hand, another negative effect is the possible partial or total wetting of the membrane pores.

He, et al. [34] did not observe any decrease of the permeate flux even for very high calcium sulfate and calcium carbonate aqueous solutions [35]. In the most unfavorable case, the decrease of the permeate flux was only 11% after 6 h of DCMD operation of a feed solution containing both calcium sulfate and calcium carbonate with the saturation index (SI) 1.21 and 49 at 75 °C, respectively. The treatment with HCl acid to reduce the pH value and limit the precipitation of carbonates was not necessary [35]. However, Gryta, et al. [68] observed a high decline of the permeate flux due to the blockage of the membrane pores when wastewater was concentrated in salts up to 48.9 g/L (mainly NaCl in presence of other salts like Mg, K and Ca) [68]. The effect of the salts crystallization on pore wetting was detected by Gryta [33,69] who found a significant wetting of pores in the presence of $CaCO_3$ crystals when treating water containing organic matter (TDS = 409–430 mg·L^{-1}, TOC = 6.8–8.5 mg·L^{-1}). This phenomenon led to a decrease of the permeate flux and an increase of the salts concentration of the permeate. Moreover, the precipitation of calcium carbonate may lead to the degassing of CO_2 that eventually is transported through the membrane pores to the permeate according to the following reactions: $2HCO_3^- \Leftrightarrow H_2O + CO_2\uparrow + CO_3^{2-}; CO_3^{2-} + Ca^{2+} \Rightarrow CaCO_3\downarrow;$ $2HCO_3^- + Ca^{2+} \overset{Heat}{\Rightarrow} H_2O + CO_2\uparrow + CaCO_3\downarrow$ [35].

When salts are precipitated, these can be removed by a simple circulation of water tangentially to the membrane [34]. He, et al. [35] cleaned the membranes by washing first with HCl (e.g., elimination of $CaCO_3$ by reducing the pH) then with pure water (e.g., elimination of salts by dissolution) followed by membrane drying. However, the mixture of different salts can form a more compact agglomerate difficult to be detached from the membrane [35].

Tun, et al. [70] and Yun, et al. [71] showed the evolution of the permeate flux with the precipitation of a mixture of Na_2SO_4 and NaCl salts. The first step is the gradual decline of the permeate flux until reaching the saturation point in which the reduction of is more significant; then a new regime corresponding to low permeate flux occurs when the membrane is almost completely covered by the salts [70,71].

Guillen-Burrieza, et al. [41] investigated the effects of membrane scaling (i.e., salt deposition) on the properties of two commercial hydrophobic membranes (i.e., PTFE and PVDF) [41]. It was proved that hydrophobic PVDF and PTFE membranes were not immune to fouling by salt deposition but behaved differently against it [41]. All the used characterization techniques to detect salt deposition showed significant scaling (i.e., presence of NaCl, $MgSiO_3$, $MgCO_3$ and $CaCO_3$) starting from the first week of seawater desalination. Figure 2 shows the cross sectional SEM images of the PTFE and PVDF membranes after the 4th week of seawater exposure. It was found that the thickness of the salt layer deposited on the membrane surface was about 7 μm for the PTFE membrane having a thickness of

50 μm (Figure 2a) and about 4 μm for the PVDF membrane having a thickness of 23 μm. Similarly, the PVDF membrane with 200 μm thickness showed a salt layer near 15 μm thick (Figure 2b) while the PVDF membrane having 125 μm thickness had about 10 μm thick salt layer [41]. In order to further understand the effects of salt deposition on MD parameters, the mechanical strength, pore size distribution and gas permeability of the membranes were evaluated before and after inorganic fouling. It was proved that inorganic fouling not only altered the membrane's properties but also the MD performance (i.e., permeate flux and salt rejection) [41].

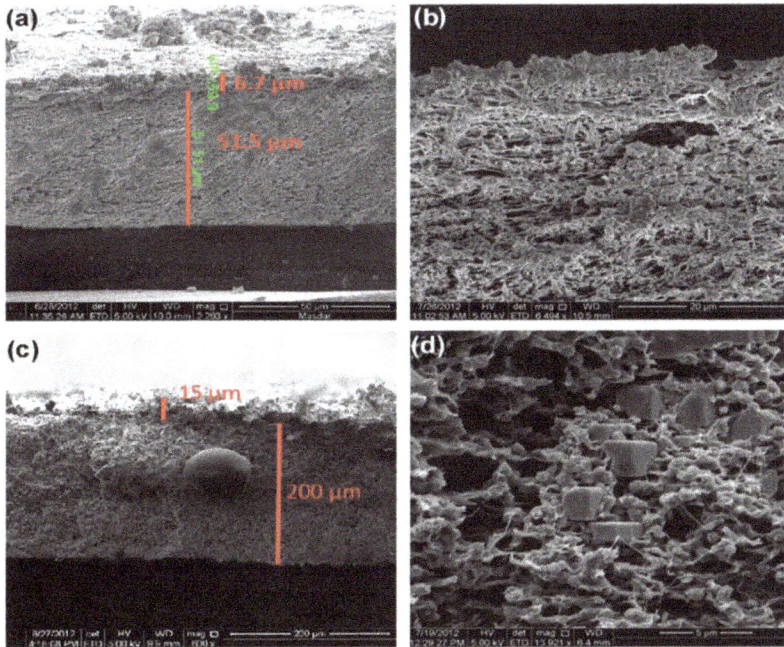

Figure 2. Cross sectional SEM images after the 4th week of seawater exposure for (**a,b**) PTFE membrane (with 0.2 μm pore size and PP nonwoven support, purchased from Thermoscientific) and (**c,d**) PVDF membrane (with 0.45 μm pore size and no support, purchased from Thermoscientific). Reproduced with permission from [41], Copyright Elsevier, 2013.

In order to prevent inorganic fouling, desalination can be carried out using a hybrid process, namely membrane crystallization, that combines both crystallization of salts and MD [70,72–75]. In this hybrid process crystallization of salts is carried out in a separate tank and the formed crystals are eliminated leading to a conversion rate very close to 100% [73].

3.1.3. Biofouling

Biofouling or biological fouling refers to the growth of bacteria or micro-organisms on the membrane surface (see Figure 3) and biological particles or colloidal species that may be trapped at both the membrane surface and/or pores forming a biofilm [69]. This type of fouling has been less studied in MD compared to other membrane processes such as MF.

It is necessary to point out that there are two types of micro-organisms: aerobic bacteria and anaerobic bacteria. For the aerobic bacteria (pseudomonas faecalis) the operating conditions such as the high temperatures applied in MD are unfavorable to its growth. However, in the case of anaerobic bacteria (*Streptococcus faecalis*) and fungi (*Aspergillus fungi*) the high temperatures applied in

MD are favorable to their reproduction and growth. In addition, *Streptococcus faecalis* can penetrate through the membrane pores [64,76]. Therefore, pretreatment by NF and addition of hypochloric acid (pH = 5) completely prevented this type fouling, even after long-term operation (i.e., after 1400 h of operation [76]).

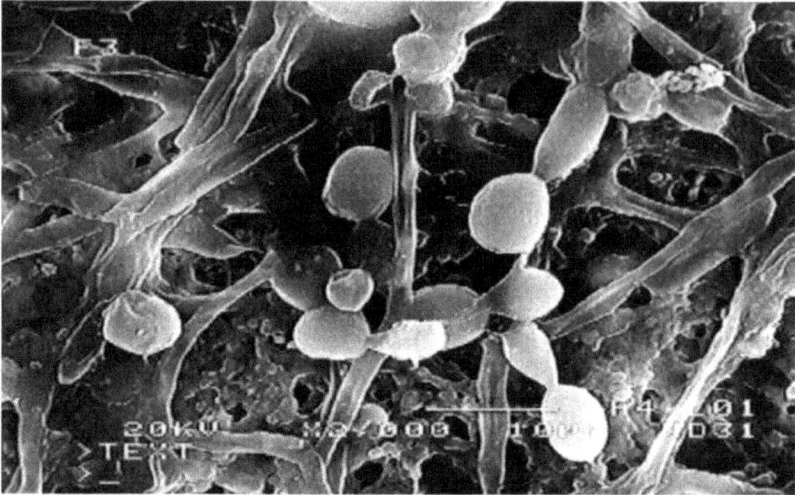

Figure 3. SEM micrograph of the micro-organisms (nonfermenting gram-negative rods) in the deposit collected from the distillate tank walls. Reproduced with permission from [69], Copyright Elsevier, 2005.

It is worth noting that EPS (i.e., extracellular polymeric substances) excreted by the bacteria are the major structural components of biofilms and are very difficult to remove [14,77,78]. The biofilm structure containing EPS with amphiphilic properties induces the membrane pore wetting leakage of feed solutes to the permeate side [15].

MD bioreactor (MDBR) consists of a MD module submerged in an aerobic–thermophilic bioreactor where the bacteria break down and consume the inorganic and organic solutes, thereby maintaining the functionality of the MD membrane module. However, due to the biomass, biofouling is a problem in MDBR as it is in conventional membrane bioreactors (MBRs). Goh, et al. [79] who used an MDBR with submerged flat sheet PVDF membranes for the treatment of synthetic wastewater, observed a permeate flux decline of 5.9% over 3 days of operation and 51% over 23 days. Confocal microscopy (CM) indicated a biofouling layer thickness of 2–8 mm after 7 days of operation and 20 mm after 22 days [79]. It was concluded that the thin biofouling layer did not significantly affect the resistance to heat transfer but it is resistant to mass transfer [3]. In another study, Goh, et al. [80] analyzed biofouling in cross-flow MD considering two sludges having different hydrophilicities. Compared to distilled water used as feed, it was observed a permeate flux decline for both sludges of 60% over 180 h. CM indicated a thickness of the biofouling layer after 180 h of 7.4–15.1 mm for the more hydrophilic sludge and 8.1–14.4 mm for the less hydrophilic one.

3.2. Fouling in OD

Similar to DCMD, OD operates under atmospheric pressure using porous and hydrophobic membranes but under the same temperature at both the feed and permeate aqueous solutions. Fouling investigation in OD has received less attention compared to MD as it can be drawn from the number of published papers summarized in Table 3.

Table 3. Published studies on membrane fouling in OD and OMD for different membrane materials and types (d_p: pore size; ε: porosity; δ: thickness; T_f: Feed temperature; T_p: Permeate temperature; v: feed velocity (or ϕ: feed flow rate); J_p: Permeate flux; FD_n: Normalized flux decline) [4,81–86].

OD/OMD	Membrane Material and Type	d_p (µm)	E (%)	δ (µm)	Foulant(s)	T_f/T_p (°C)	v (m·s⁻¹)	J_p (kg·m⁻²·h⁻¹)	FD_n (%)	Observations	Ref.
OD	PTFE Flat-sheet	0.2 0.45 1	80	178	Phenolic compounds from crude olive mill wastewater	30	500 rpm	2.64–2.23 3.01–2.64 3.86–2.85	15% 13% 26%	The decrease of the permeate flux with time is more obvious for the membrane which pore size of 1 µm.	[81]
		0.2	78	8.5	Tomato puree	20–24	ϕ: 0.5 L·min⁻¹	1.25 for 6 wt % 0.7 for 21.5 wt %	-	Tomatoes are composed of 95% water, 3–4% carbohydrate, 0.51% protein and 0.1–0.3% fat.	[82]
		0.2 0.45 1.2	-	-	Phenolic content of red grape juice	35	-	8–4	84%	Initial juice concentration 5 °Brix. Initial concentration of stripping solution 50 wt % CaCl₂.	[83]
	PVDF Hollow fiber	0.2	75	125	Glucose	25, 35, 45	0.4; 0.6; 0.8	1.67 to 4.73 Concentration factor: From 30 to 40 °Brix	-	Feed concentration and brine velocity have significant effect on OD permeate flux (while the brine concentration was remained constant) Their effects depend on the range of the feed concentration.	[4]
		0.2	64	170	Glucose	25, 35, 45	0.2; 0.4; 0.5	1.00 to 2.87 Concentration factor: From 45 to 60 °Brix	-		
	PP	0.2	-	125	Apple juice	23–33	ϕ: 10 L·min⁻¹	From 2.25 to 0.9	71%	OD can be almost free from fouling due to the hydrophobic OD membrane. Few substances (e.g., fat and wax) may stick to the membrane surface.	[84]
OMD	PTFE Flat-sheet	0.2	80	178	Phenolic compounds from crude olive mill wastewater	40/20	500 rpm	From 3.9 to 1.3	13.5%	Membrane fouling is of less importance when using OMD by PTFE membranes.	[81]
	PP Hollow fiber	0.04	40	40	Crystals formed from Na₂CO₃	40/20	ϕ: 1.4, 0.6, 0.45 L·min⁻¹	From 0.12 to 0.078	30%	Membrane scaling was observed is due to the accumulation of crystals on the membrane surface.	[86]

As stated previously, polarization phenomena exert a major influence on fouling and scaling. According to Bui, et al. [87], who quantified the effects of the concentration and temperature polarization in OD using glucose solutions and PVDF hollow fiber membranes, found that the polarization phenomena contributed up to 18% of the permeate flux reduction. In OD, provided that both sides of the membranes are brought in contact with the feed aqueous solution to be treated and the osmotic solution in the permeate side, polarization effects are more significant than in DCMD [87]. In a similar study, Bui and Nguyen [84] used PP membranes to concentrate an apple juice solution and observed a permeate flux decline from 2.25 to 0.9 kg/m²h. This was explained by the grip of some substances (e.g., fat and wax) to the membrane surface. Cleaning using filtered water and NaOH (0.1–1%) aqueous solution in the feed side of the OD system was carried out to recover the initial permeate flux.

Durham and Nguyen [82] used Gore-tex PTFE membrane and Gelman 11104/2TPR (a cross linked acrylic fluorourethane copolymer) to concentrate tomato puree by OD and to study membrane fouling due to the adhesion of fatty components, tomato pigments, lycopene and beta carotene to the membrane surface. Tomatoes are composed of 95% water, 3–4% carbohydrate, 0.51% protein and 0.1–0.3% fat [88,89]. It was found that the permeate flux decreased from 1.25 kg/m²h when the feed solution contained 6% of Tomato to 0.7 kg/m²h when the feed contained 21.5% of tomato. The repetitive fouling and cleaning of the membrane with either water, P3 Ultrasil 56 or 1% NaOH resulted in membrane pore wetting allowing salt leakage into the feed after only 2 to 3 cleaning runs [82]. The authors claimed that Gelman 11104/2TPR membrane was more suitable than Gore-tex PTFE membrane for the concentration of tomato puree by OD [82].

El-Abbassi, et al. [81] evaluated PTFE membranes fouling in OD by comparing the measured water permeate flux before and after crude olive mill waste water (OMW) treatment. This was expressed by means of the permeate flux reduction rate (*FR*) defined as [81]:

$$FR(\%) = \left(1 - \frac{D_{WFa}}{D_{WFb}}\right) \times 100 \tag{2}$$

where D_{WFa} and D_{WFb} are respectively the water permeate flux after and before OD of OMW treatment under the same operating conditions.

It was found that *FR* depended on the membrane pore size (i.e., 2.94, 4.02 and 4.14% for the membranes TF200 (0.2 μm pore size), TF450 (0.45 μm pore size) and TF1000 (1 μm pore size), respectively). However, in all cases all the used PTFE membranes showed a high fouling resistance and *FR* did not exceed 5%. The decrease of the permeate flux with time was more obvious for the membrane TF1000 that exhibited the highest permeate flux (i.e., the permeate flux decreased by 15%, 13% and 26% for the membranes TF200, TF450 and TF1000, respectively; after 280 min of OMW treatment by OD) [81].

Kujawski, et al. [83] applied OD for the dehydration of red grape juice of different concentrations (5–20 °Brix) at 35 °C using three PTFE membranes with different pore sizes (0.2, 0.45 and 1.2 μm) and calcium chloride CaCl₂ solution (50 wt %) as stripping solutions. The permeate fluxes of 5 °Brix red grape juice decreased from ≈8 to ≈4 kg/m²h during 550 min of operating time and the permeate fluxes of the membranes having smaller pore size (0.2 μm) were slightly higher due to less surface fouling and pore blockage. The suspended particles with a size bigger than 0.2 μm can adhere easily and block the membrane pores with 0.45 and 1.2 μm size.

3.3. Fouling in OMD

Fouling in OMD may occur following the same mechanisms mentioned previously in DCMD and OD fouling, provided that OMD is a non-isothermal process using an osmotic solution in the permeate side of a porous and hydrophobic membrane. Similar to any other membrane process, the presence of fouling in OMD may vary according to the nature of the feed solution to be treated and the characteristics of the used membrane. It is worth noting that fouling investigation in OMD has received little attention as can be concluded from the number of published papers summarized in Table 3. It was reported that fouling in OMD comprises a major obstacle for its industrial implementation and the determination of the OMD performance during long-term run is required [7].

El-Abbassi, et al. [81] used the OMD process to treat OMW with PTFE membranes of different pore sizes (TF200, TF450 and TF1000) at 40 °C feed temperature and 20 °C of 5 M CaCl$_2$ osmotic permeate solution. To study membrane fouling, OMD experiments were carried out using distilled water as feed before and after each OMW treatment [81]. Long-term experiments of 30 h OMW processing were performed. The permeate flux showed a decrease from 3.9 to 1.3 L/m^{2}h (i.e., a decline of 67%). However, after rinsing the fouled membrane with distilled water, a reduction of only 5.7% of the permeate flux of water was detected indicating that membrane fouling is of less importance when using OMD for the treatment of OMW by PTFE membranes [81].

In a recent study, Ruiz Salmón, et al. [86] conducted OMD-crystallization to obtain Na$_2$CO$_3 \cdot$10H$_2$O as a solid product using NaCl as an osmotic solution and a feed solution containing Na$_2$CO$_3$. A hollow fiber PP membrane with an effective pore size of 0.04 μm and a porosity of 40% was employed. The feed temperature ranged between 20 and 40 °C while that of permeate was kept at room temperature. Different experiments were carried out varying the concentration of Na$_2$CO$_3$ and NaCl in the feed and permeate, respectively. A decrease of the permeate flux was detected during the first minute of each experiment, and scaling was observed in some experiments due to the accumulation of crystals on the membrane surface, coming from the feed reservoir because of the recirculation of the feed stream [86].

4. Characterization Techniques for Fouling Analysis

Fouling affects directly both the hydrophobic membrane surface and its pores reducing the MD, OD and OMD separation performances. In order to understand fouling mechanisms in these processes and mitigate therefore their effects chemical and structural characterization must be performed using different techniques. Some of them are currently used in MD, but to a lesser extent than in OD and OMD. Other techniques are not used yet in any of these processes, but are cited for the sake of recommendation provided that they have been considered in other membrane separation processes such as RO, NF and UF. The following characterization techniques include those that permit us to figure out the presence or absence of fouling, thickness of fouling layer and its effect on membrane morphological characteristics; and those that allow us to quantify different fouling components.

4.1. Visual Analysis

The fouled membrane can be initially inspected using a light microscope equipped with a digital camera focused on the feed side of the fouled membrane. This visual analysis was successfully used for flat sheet membranes in order to observe in situ particles deposition with time. With this technique it is impossible to visualize neither the thickness of the fouling layer nor the fouling composition, and only particles whose size is greater than 1 μm can be detected [15].

4.2. Microscopy Techniques

Several electronic microscopy techniques can be used to characterize the morphological structure of porous membranes (top and bottom surfaces for flat sheet membranes, internal and external surfaces of capillaries and hollow fiber membranes as well as their cross-sections). These techniques include scanning electron microscopy (SEM), transmission electron microscopy (TEM), field emission scanning electron microscopy (FESEM), atomic force microscopy (AFM), etc. By means of these techniques, various membrane parameters can be determined such as the membrane mean pore size, pore density, pore size distribution, surface porosity, roughness, fouling particles size and thickness of the fouling layer. These characteristics permit us to detect the presence or absence of foulants on the membrane surface.

4.2.1. Scanning Electron Microscopy (SEM) and Energy Dispersive X-ray Spectroscopy (EDX)

SEM is one of the most used techniques to study both membrane surface and its cross-section [90–93]. In this technique, with the electron/sample interactions and the reflected electrons high resolution topographical images of fouled membranes are provided [94]. First, liquid nitrogen is used to freeze the membrane sample that is subsequently broken in small pieces [2,15]. Then the sample is coated with a layer of gold, carbon or platinum by sputtering to render it electrically conductive [15,95]. It must be pointed out that due to both immersion in liquid nitrogen and coating, the fouled membrane sample limits the use of this technique for fouling characterization because it gives the sample some artifacts and damages changing the fouling layer characteristics [2,15,96]. More details on this technique can be found in [2,15].

In addition, energy dispersive X-ray (EDX) spectroscopy can be applied together with SEM to analyze the composition and crystallographic nature of the membrane sample [94]. The principle of this technique relies on the interaction of a source of X-ray excitation and the metal coated sample. In this case, when an incident electron bombards an atom of the sample and knocks out an electron from the outer layer of the metal coated sample, a vacancy or hole is left in this layer. If an electron from another layer fills this vacancy (electron transitions), then X-rays are emitted. These transitions are characteristic of each chemical element [94].

4.2.2. Transmission Electron Microscopy (TEM)

This microscopy technique may be considered complementary to SEM mainly when the fouling structures are too small to be detected by SEM [97,98]. It is also adequate to study the presence of fouling particles inside the membrane pores. In this technique, a sufficiently accelerated electron beam collides with a thin sample (i.e., about ten nanometers). Depending on the sample thickness and the type of atoms forming it, some electrons cross the sample directly and others pass through it but are diverted. After passing the sample the electrons are collected and focused by a lens to form a final image on a CCD camera with a high definition. If the image is formed from the transmitted beam, which has not undergone scattering, then the image of the object is dark on a bright background. On the other hand, if the scattered electrons are used in this case, the image appears bright on a dark background. Therefore, these two techniques are called image formation in a clear field and in a dark field, respectively; the first one is the most used. More details on this technique can be found in [99].

Among the followed procedures to prepare transparent specimens to electrons, the most important and most used one is based on a mechanical thinning of the material in a very controlled way. This produces a flat specimen of a few microns thick with a flawless surface, which is then subjected to ionic surface polishing at low angle and low energy to achieve extremely thin areas ready for TEM observation. Ultramicrotomy is another technique used to produce TEM specimens, usually reserved for soft materials. The procedure consists of cutting sample slices using diamond blade. Polymeric samples of about 50 nm thickness can be prepared. Samples can be cut in a temperature range between $-180\,^\circ$C and ambient temperature depending on the characteristics of the material.

It is worth noting that TEM is not applied yet in MD, OD and OMD fouling analysis due to the necessity to prepare extremely thin samples to be transparent to electrons affecting considerably the fouling layer.

4.2.3. Atomic Force Microscopy (AFM)

AFM is one of the advanced techniques used for characterization of the membrane surface without requiring any previous sample preparation. Unlike conventional optical or electron microscopy, AFM requires physical interaction between a probe with a sharp tip and the sample surface. It is possible to work in contact, non-contact or intermittent contact mode to get the three-dimensional topographical images of the membrane surface. The measurements can be made in different media: air, liquid, vacuum, controlled atmosphere, etc. The AFM probe scans the sample onto which a laser beam is directed and reflected onto a four-quadrant photodiode. The feedback controller evaluates the signal coming out of the photodiode. Moreover, piezoelectric elements are included in the scan head to ensure precise nanoscale motion [100]. Comprehensive information on this technique can be found in [101,102].

In general, AFM technique is used to determine surface roughness parameters, mean pore size, pore size distribution, surface porosity and pore density of both new and fouled membranes [101]. This technique is commonly used for characterization of MD membranes. As an example, Zarebska, et al. [103] used AFM morphological analysis to understand the observed decrease in contact angle and surface charge of PTFE and PP membranes used in MD. The three-dimensional AFM pictures of virgin and fouled PTFE and PP membranes with model manure solution and pig manure after sieving with MF and UF showed that the foulants accumulate in the "valleys" leading to a drop of surface roughness [103].

4.2.4. Confocal Laser Scanning Microscopy (CLSM)

CLSM technique is used to analyze both the surface and internal structure of the membrane based on the fluorescence emitted by a sample after its irradiation by a laser beam, obtaining three dimensional images of the samples. It permits us to increase the optical resolution and contrast by means of eliminating the out-of-focus light, achieving a controlled and highly limited depth of focus. CLSM is a non-destructive technique allowing in situ visualization of membranes and one of the most important techniques in the field of fluorescence imaging and scanning to obtain high-resolution optical images [15,104].

CLSM is basically used to take images of membrane biofouling, combining the laser scanning method with the 3D detection of biological objects labeled with fluorescent markers. It runs by focusing a laser beam onto a small sample of the fouled membrane and then the reflected light is detected by a photodetection device. The images are then acquired point-by-point and reconstructed with a computer, allowing three-dimensional reconstructions of topologically complex objects [15,104]. More details on this technique can be found in [105,106].

It is worth noting that few authors used this technique to characterize the membrane biofouling. Yuan, et al. [106] used CLSM to observe the growth of biofouling layer in osmotic membrane bioreactors (OMBRs). The CLSM images showed that during OMBRs operation the variation of the quantity and distribution of polysaccharides, proteins and microorganisms in the biofouling layer were significantly different. Ferrando, et al. [107] used CLSM to determine the fouling produced during the filtration of protein solutions. Other authors considered this technique to validate the efficiency of membrane cleaning [108,109]. However, this technique is not applied yet in MD, OD and OMD fouling analysis.

4.3. Contact Angle

In general, contact angle measurement is considered to quantify the hydrophobic character of the membranes used in MD, OD and OMD. It can be used to determine the hydrophobicity reduction of the membrane caused by fouling. To carry out this measurement, the sample does not require any previous preparation. The tip of a syringe is placed near the sample surface and then depressed so that a constant liquid drop volume (i.e., distilled water) of about 2 μL is deposited on it. Images of the drop are taken by a camera and a specific software permits to determine the liquid contact angles. A mean value together with its standard deviation are finally calculated from more than ten to fifteen readings [2]. In addition to the water contact angle, the same system can be used to measure the surface tension of the membrane and then figure out its change depending on the adsorbed foulants compared to new membrane. More details on this technique can be found in [2,110].

Various authors have used this membrane surface characterization technique before and after carrying out the separation process. According to Guillen-Burrieza, et al. [111] a clear decrease of the water contact angle value from an average of 129° for the virgin PTFE membrane to 108° for the fouled one (i.e., scaling) was observed. Sanmartino, et al. [112] also detected smaller water contact angles of fouled PTFE membranes (TF200 and TF450) than those of new ones. The membrane having higher values of crystallization fouling factor exhibited the lower value of the water contact angle. Zarebska, et al. [58] measured the water contact angles of new and fouled PP membranes used in MD. Figure 4 shows the water drops on the surface of these membranes. A clear water contact angle reduction from 142° for the clean PP membrane to 91° for the fouled membrane. It was claimed that the loss of the membrane hydrophobicity was attributed to partial wetting of the membrane due to fouling deposit (i.e., inorganic ions, proteins and/or peptides and microorganisms present in pig manure).

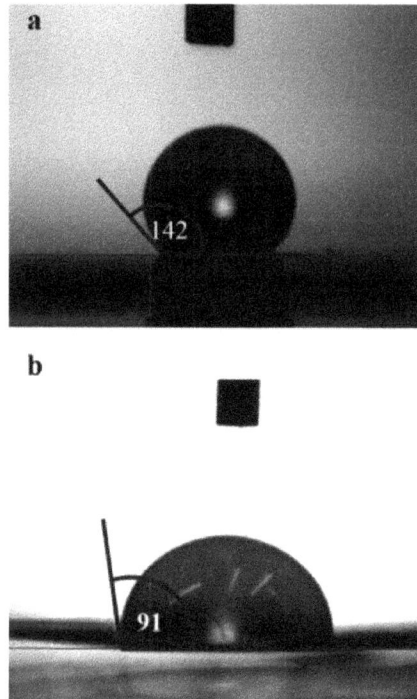

Figure 4. (**a**) Water droplet on clean flat sheet polypropylene (PP) membrane; (**b**) Water droplet on flat sheet PP membrane fouled by pig manure. Reproduced with permission from [58], Copyright Elsevier, 2014.

4.4. Infrared Thermography Technique (IRT)

IRT is a non-contact method that permits, through the infrared radiation emitted by objects, the measurement of the surface temperature and its distribution [113]. Before using this technique, the membrane sample must be dried, assuming that the temperature is affected by the humidity, and the temperature on the membrane surface must be determined correctly. In this method, an IR camera obtains infrared images due to IR radiation coming from the fouled membrane [114]. It permits the measurements of the emissivity of foulant(s) and membrane surface temperature. More details on this technique can be found in [113,115].

The aim of using this technique is to distinguish between foulants having metallic properties from those that are non-metallic, as can be seen in Figure 5 [115]. The results obtained with this technique were compared to those obtained with SEM-EDX analysis to corroborate the ability of the IRT technique to figure out whether a foulant was metallic or non-metallic in nature. Ndukaife, et al. [115] used IRT technique to study fouling of an UF flat sheet membrane. Different fouling experiments were realized using an aqueous feed solution containing an aluminum oxide nanoparticle. The results showed that the technique could detect the modifications that occur on the membrane surface after desalination process due to foulants deposition. The emissivity of the membrane surface depended on the fouled surface roughness and the composition of the foulant(s) [115]. It is to be noted that no fouling analysis was carried out yet with IRT in MD, OD and OMD processes.

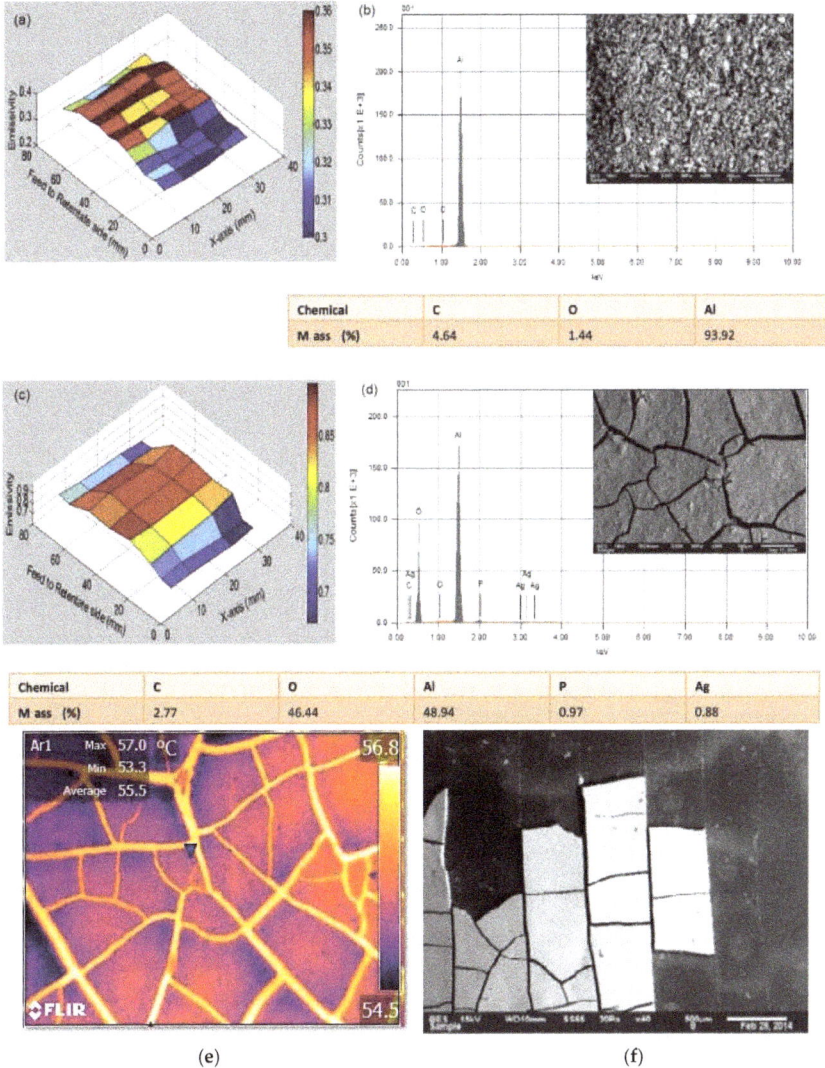

Chemical	C	O	Al
M ass (%)	4.64	1.44	93.92

Chemical	C	O	Al	P	Ag
M ass (%)	2.77	46.44	48.94	0.97	0.88

(e)   (f)

Figure 5. 3D plots showing emissivity values of various locations on the fouled membranes at 1833 ppm feed concentrations of the synthetic feed water alongside SEM images and EDX analysis (**a**,**b**) aluminum; and (**c**,**d**) aluminum oxide. Comparison of images obtained by (**e**) IRT and (**f**) SEM. Adapted with permission from [115], Copyright Elsevier, 2015.

4.5. Ultrasonic Time Domain Reflectometry (UTDR)

UTDR is a non-invasive technique used to monitor the deposition of combined organic and colloidal fouling on membrane surface [116]. This does not require any sample preparation and is based on the propagation of mechanical waves. The reflection and transmission can occur when an ultrasonic wave encounters an interface between two media. More details can be found in [116–118]. UTDR technique coupled with differential signal analysis was used to investigate the combined fouling deposition processes on membrane surface [118]. The steady-state waveform of distilled water was considered as the reference and the waveform of fouled membrane was considered as the test waveform. The difference between the two signals represented the signal of the fouling layer. UTDR was used for fouling detection by different authors in filtration processes. As an example, Li, et al. [116] investigated the fouling behavior of mixed silica–BSA solution and silica–NaCl solution using UTDR technique in NF membranes (see Figure 6). The UTDR results of NF experiments showed that the fouling layer obtained from the combined organic–colloidal fouling (mixed silica and bovine serum albumin (BSA) and silica and NaCl solution) was denser than that obtained from the colloidal fouling layer in the presence of NaCl. The formation of the denser fouling layer was due to the electrostatic interactions between foulants and membrane surface as well as the electrostatic interactions among the foulants due to absorption of BSA onto silica [116]. Taheri, et al. [119] used UTDR technique to provide a good estimation of the fouling layer thickness formed during RO tests. Xu, et al. [120] described the extension of UTDR as a non-invasive fouling monitoring technique used for the real-time measurement of particle deposition in a single hollow fiber membrane during MF. Tung, et al. [121] used a high-frequency 50-MHz ultrasound system to measure the fouling distribution in spiral wound UF and RO membrane modules. It is necessary to point out that this technique has not been applied yet in MD, OD and OMD fouling.

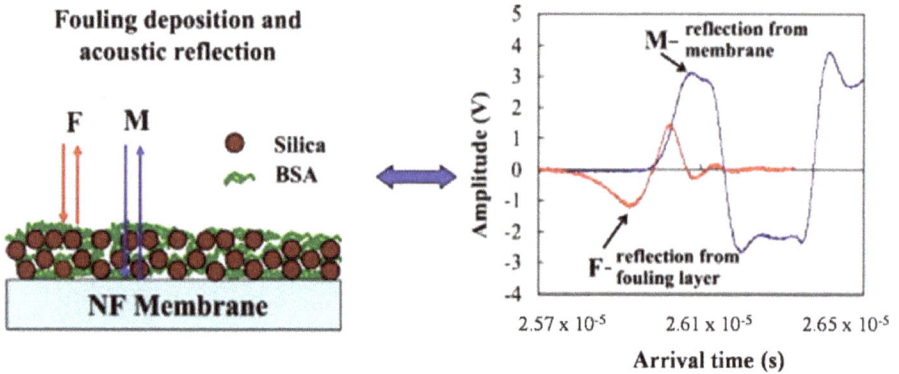

Figure 6. Ultrasonic signal responses of the clean NF membrane (blue line) and differential signals (red line) obtained at 300 min of fouling operation with the feed solution of 1000 mg/L silica and 1000 mg/L NaCl under a constant pressure operation. Adapted with permission from [116], Copyright Elsevier, 2015.

4.6. Zeta Potential

Fouling can change the membrane surface charge and by measuring the zeta potential of new and fouled membranes one can evaluate the possibility of foulant(s) to adhere or not to the membrane surface. It is a measure of the magnitude of electrostatic repulsion or attraction between particles. In general, zeta potential is a key indicator of the stability of colloidal particles and it is a good method to describe double-layer properties of colloidal dispersions [15,103]. Various factors affect the zeta potential such as the concentration, composition, temperature and pH of the solution as well as the membrane surface properties (i.e., charge, roughness, chemical heterogeneity, etc.) [122]. Therefore, it is a useful technique to determine whether a membrane is modified in order to improve the considered process performance. The determination of the zeta potential can be carried out using several methods such as the streaming potential, vibrational potential, electrophoresis, electro-osmosis, etc. For fouling characterizations streaming potential and electrophoresis are the most considered methods [123].

Electrophoresis is the movement of charged particles or polyelectrolytes in a liquid under the influence of an external electric field. The electrophoretic mobility of a charged particle is determined by the balance of electrical and viscous forces that could be used to determine the zeta potential. The streaming potential is measured when the electrolyte solution is forced with a known pressure through a channel [124]. A resulting voltage is measured between electrode probes on either side of the channel and compared with the voltage at zero applied pressure. The streaming potential can be related to the zeta potential by factors that include the electrical conductivity, fluid viscosity and the structure of the channel [123]. More experimental details of zeta potential can be found in [122].

Zarebska, et al. [103] measured the streaming zeta potential of PP and PTFE membranes after their use in MD for the treatment of different pretreated model manure feed solutions. It was stated that new PP and PTFE exhibited a small negative zeta potential (above −30 mV, at pH ≈ 9). The fouled PTFE membrane exhibited the more negative zeta potential (above −40 mV, at pH ≈ 9) indicating the higher electrostatic repulsion between model manure solution and the membrane surface. However, for the PP after adsorptive fouling with model manure solution the zeta potential was above −15 mV (at pH ≈ 9) (see Figure 7) suggesting an increased fouling potential due to decreased electrostatic repulsion between the membrane surface and the model manure solution.

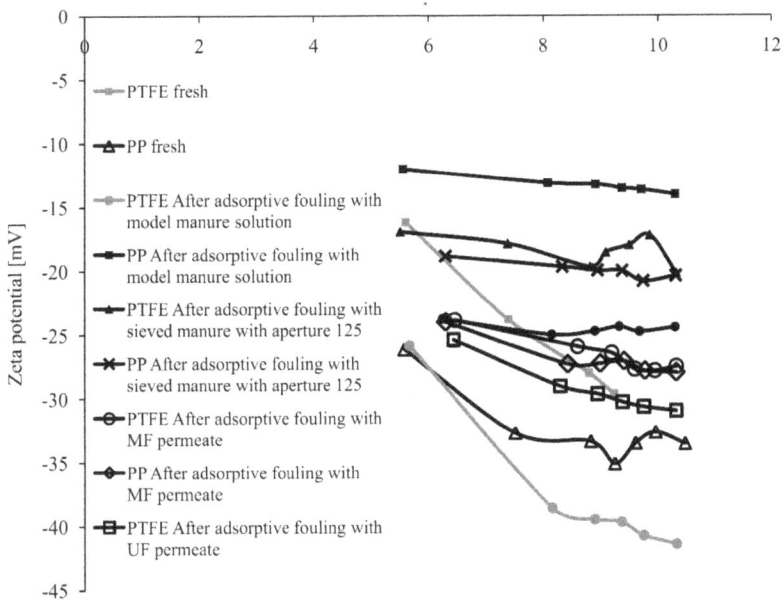

Figure 7. Zeta potential of polytetrafluoroethylene (PTFE) and PP membranes as a function of pH after adsorptive fouling. Reprinted with permission from [103], Copyright Elsevier, 2015.

An et al. [29,30] characterized both commercial and prepared MD membranes by means of zeta potential at different pH values and claimed that the negatively charged MD membranes were resistant to dye adsorption and subsequent fouling.

4.7. X-ray Diffraction (XRD)

XRD is an analytical measurement technique that identifies the crystalline nature of the foulant(s) either organic or inorganic, polymeric or metallic deposited on the membrane surface. It reveals important information about the crystal structure, size, shape, etc. Different type of polycrystalline samples can be measured. The foulants deposited on the membrane surface can be extracted and ground to a powder form, and then arranged in a very thin layer on a sample holder. On the other hand, membrane samples can also be measured directly. Both types of samples are usually analyzed with a diffractometer equipped with monochromatic Cu-Kα radiation [95]. The irradiated crystals disperse the X-rays only in some determined directions with intensities that depend on how the atoms are ordered at the microscopic level. With this information, direction and intensity of each ray, it is able to obtain the molecular structure of crystals. More details on XRD technique and analysis can be found in [125].

It is necessary to point out that XRD has widely been applied to study membrane fouling [15], even in MD. As example, Gryta [46] used XRD to analyze the components of scaling on PP membrane. As can be seen in Figure 8, the spectra showed the presence of calcite crystals when the antiscalant was not used, and the calcite peak disappeared after using the antiscalant. Kim, et al. [126] characterized the components making up the crystals formed in the crystallizer of an integrated MD and membrane crystallization (MDC) system for shale gas produced water (SGPW) treatment.

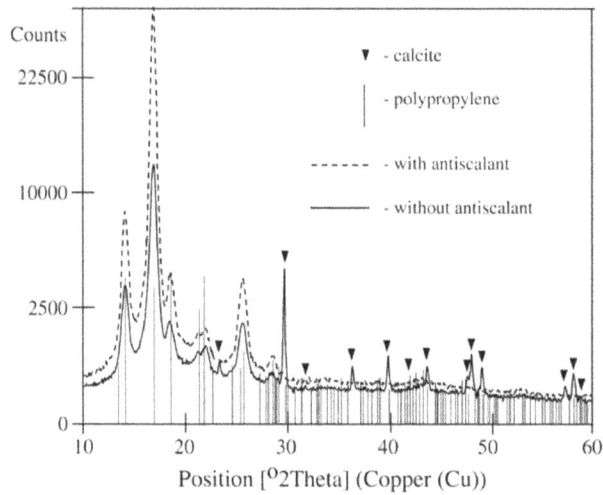

Figure 8. X-ray Diffraction (XRD) analysis of the deposit formed on the membrane surface. Broken line—without antiscalant, continuous line—with 10 ppm antiscalant. Feed solution prepared by dissolving NaHCO$_3$ and CaCl$_2$ (mole ratio 2:1) 3.1 mmol HCO$_3^-$/dm^3. Reprinted with permission from [46], Copyright Elsevier, 2012.

4.8. X-ray Fluorescence (XRF)

XRF is a semi-quantitative analytical technique based on wavelength-dispersive spectroscopic principle similar to an electron microprobe [127,128]. The sample material emits secondary or fluorescent X-rays after being excited by a primary X-ray source. Each of the elements present in the sample produces a set of characteristic fluorescent X-rays (i.e., a fingerprint) that is unique for that specific element, which is why XRF spectroscopy is an excellent technology for elemental and chemical analysis. This technique requires a previous sample preparation like XRD.

The utility of this technique in membrane fouling characterization is to determine the elemental composition of both new and fouled membranes and subsequently figure out the chemical nature of the deposit(s). This technique is distinguished by its highest accuracy and precision as well as by its simple and fast sample preparation for the elemental analysis. More details of sample preparation and procedure concerning this characterization technique can be found in [127,128]. It must be mentioned that this technique is not used yet for fouling analysis in MD, OD and OMD. However, it was used in the assessment of long-term fouling of RO membranes [127].

4.9. Attenuated Total Reflectance-Fourier Transform Infrared Spectroscopy (ATR-FTIR)

ATR-FTIR offers quantitative and qualitative analysis of both organic and inorganic solid and liquid samples and it is a well-known technique for fouling analysis.

It is a label free technique that provides information on the chemical composition of the fouling layer [129]. ATR-FTIR measurements only collect signals from the surface and to a depth of 1.6 mm inside the sample by using a beam splitter KBr and an infrared source employing an attenuated total reflectance. When the infrared radiation reaches the sample, part of the radiation is absorbed by the sample and the rest is transmitted through it. The resulting signal in the detector is a spectrum representing the molecular "fingerprint" of the sample. The utility of infrared spectroscopy is the fact that different chemical structures (molecules) produce different spectral traces. The Fourier transform converts the output of the detector into an interpretable spectrum that provides information about chemical bonds and functional group in a molecule. In this technique the sample does not need any specific type of preparation. Further details about this technique can be found in [103,129].

Thygesen, et al. [129] used ATR-FTIR technique to determine the composition of the fouling layers deposited on PP and PTFE membranes used for ammonia removal from model manure by MD. It was proved that the fouling layer was formed by only organic compounds, whereas no indication of inorganic scaling was detected [129]. Zarebska, et al. [103] used ATR-FTIR to detect functional groups in PP and PTFE membranes used in MD for the treatment of model manure solution. This technique showed that fouling layer composition was not uniform across the entire membrane surface and all the bands found by ATR-FTIR spectroscopy were assigned to organic fouling. Tomaszewska and Białończyk [130] also used ATR-FTIR method to analyze fouling composition during whey concentration process by MD. The ATR-FTIR spectra of both clean and fouled PP membrane confirmed the presence of whey proteins in the fouled membrane structure.

4.10. Inductively Coupled Plasma Atomic Emission Spectrometry (ICP-AES)

ICP-AES is also referred as inductively coupled plasma optical emission spectrometry (ICP-OES). It is a type of emission spectroscopy that uses the inductively coupled plasma to produce excited atoms and ions that emit electromagnetic radiation at wavelengths characteristic of a particular element. It is a flame technique with a flame temperature in the range 6000–10,000 K. The intensity of this emission is indicative of the concentration of the element within the sample. It is used for the detection of trace metals (e.g., Al, Cu, Fe, Cr, Zn, Ni, B, Mn, etc.) [123,127], being very suitable for inorganic fouling but it gives only qualitative information. Consequently, it needs to be coupled with additional analytical methods like atomic absorption spectroscopy (AAS) mentioned below. ICP-OES technique requires an exhaustive preparation of the sample before its analysis. This is one of the reasons why it is not a very used technique in fouling characterization. Very few studies have been conducted using this technique. For example, Nguyen, et al. [131] used ICP-OES to analyze the inorganic elements accumulated in the fouling layer of MD membrane used with SWRO brines. A possible sample preparation is the immersion in an acid solution such as HCl or H_2SO_4 in order to dissolve the precipitates and transform them into ionic form or heating the sample in furnace at 550 °C for about 16 h followed by a dissolution in an acid solution (HCl or HNO_3) and the subsequent analysis of the obtained solution(s) [123]. ICP-OES is very suitable for inorganic fouling. More details can be found in [123,127].

4.11. Atomic Absorption Spectroscopy (AAS) Analysis

AAS is a spectroanalytical procedure employed for the quantitative determination of chemical elements in a sample using the absorption of optical radiation (i.e., light) by free atoms in the gaseous state. It is based on the analyte atomization using different atomizers in a liquid matrix. The electrons of the atoms in the atomizer can be promoted to higher orbitals (excited state) for a short period of time (nano-seconds) by absorbing a defined quantity of energy (radiation of a given wavelength). This amount of energy (i.e., wavelength) is specific to a particular electron transition in a particular element. In general, each wavelength corresponds to only one element, and the width of an absorption line is only of the order of a few pico-meters, which gives the technique its elemental selectivity.

The determination of fouling composition by AAS needs the same sample preparation as that indicated for ICP-OES [123,127]. As it is explained above, for fouling analysis it is usually coupled with ICP-OES. As an example, Melián-Martel, et al. [127] used this method in the determination of the long-term fouling of RO membranes.

4.12. Excitation Emission-Matrix Fluorescence Spectroscopy (EEM)

EEM technique has been widely used to characterize dissolved organic matter in water and soil. It is based on the fluorescence analysis from a sample by using a luminescence spectrometer. A beam of light, usually ultraviolet light, excites the electrons in molecules of certain compounds and causes them to emit light. In the field of membrane science, it is useful for the detection of HA and other organic matter present in fouled membranes. The use of EEM technique demonstrated that changes in a sample, containing HA, proteins, etc., can considerably change the intensity distribution of the fluorescence spectra, particularly if strong adsorbant non-fluorescent species are present [132]. Solid sample slices are directly mounted in the sample compartment of the spectrofluorimeter. More details on sample preparation and method can be found in [132,133]. To date there has been no MD, OD or OMD research study in which this technique has been applied.

4.13. Field Flow Fractionation (FFF)

FFF is a technique applied for the separation and characterization of macromolecules, supramolecular assemblies, colloids and particles. In this separation technique, a field is applied perpendicularly to the laminar flow of a carrier liquid in order to cause the separation of the particles present in this liquid, depending on their different mobilities under the force exerted by the field. The particles change their positions (i.e., levels) and speed depending on their size/mass. Since these components travel at different speeds, their separation occurs. This separation can be carried out with a high resolution over a wide size range from 1 nm to 100 μm [134].

Various fields can be applied in FFF depending on the nature of the material to be analyzed including flow field-flow fractionation (FlF-FF), thermal (ThFFF), electrical (ElFFF), sedimentation (SdFFF), gravitational (GrFFF), dielectrophoretic (DEP-FFF), acoustic (AcFFF) and magnetic (MgFFF) field-flow fractionation [134]. Further details on these techniques can be found elsewhere [134,135].

It must be pointed out that FlF-FF is the most commonly used FFF techniques for surface and polymer analysis. One of the largest areas of active research in FlF-FF is in the area of proteins, bacteria and sub-cellular structures [134]. Therefore, Fl-FFF can be effectively used to predict biofouling [136]. However, this technique is still not considered in MD, OD or OMD fouling analysis.

5. Methodologies for Membrane Fouling Prevention and Reduction

Different strategies have been adopted to improve the performance of MD, OD and OMD processes in terms of fouling prevention and extension of the life-time of membranes. These can be established in two possible ways: (i)—to treat the feed solution by means of pretreatments to diminish the foulants content or by adding antiscalants to inhibit inorganic scaling during the process; (ii)—to design fouling resistant membranes with improved MD, OD or OMD performance considering membrane modification or to establish systematic membrane cleaning steps during MD, OD and OMD processes.

5.1. Pretreatment

Appropriate and effective pretreatment is one of the essential keys to improve MD, OD or OMD processes and minimize their fouling and scaling problems. To date, pretreatment has been claimed to be the effective method to prevent and control fouling in the MD process [15]. The degree of pretreatment depends on the type of membrane material, nature of the feed solution, water recovery level and frequency of membrane cleaning [137,138]. Pretreatment can change the properties (chemical and/or biological) of the feed solution leading to less fouling formation and improving the permeate flux, separation factor and life-time of the membrane. Pretreatment techniques and technologies can be categorized in three types: (i) mechanical; (ii) chemical; (iii) thermal and (iv) combination of mechanical and chemical or mechanical and thermal procedures. Table 4 summarizes some pretreatments considered before carrying out MD and OD processes.

Mechanical pretreatment involves physical techniques such as membrane-based filtration (NF, UF or MF) to reduce suspended solids, colloids, microorganism and organic matter present in feed aqueous solutions subjected to MD, OD or OMD treatment. It is important to reduce or eliminate these species previous MD, OD and OMD process in order to minimize membrane fouling and prevent damage of the membrane. NF and RO permeates after treatment of surface water were used as feed for MD process, the results showed that the fouling problem was reduced in the MD process [139]. In spite of NF permeate, which still contained a significant amount of carbonate that can form a fouling layer on the membrane surface, small amounts of HCl (pH = 5) added to the NF permeate permitted the elimination of this phenomenon in the MD process. When the RO permeate was employed as feed for the MD process, the problem of fouling was not detected.

Bailey, et al. [140] studied the effect of UF as a pretreatment on the subsequent concentration of grape juice by OD. UF using membranes with pore diameters of 0.1 μm or less showed an important enhancement of the permeate flux compared to that observed without using UF pretreatment. Furthermore, it was found that the content of fermentable sugars, considered as glucose and fructose, of whole juice was decreased and the removal of proteins with UF resulted in an enhancement of the juice surface tension and the subsequent reduction of the risk of membrane pore wetting.

Table 4. Pretreatments considered in MD and OD processes.

Pretreatment	MD/OD Set Up	Feed Solution	Species Addressed	Observation	Ref.
UF, NF, RO	DCMD	Surface water (lake)	Organic compounds, suspended solids, colloids	- UF reduced the salt density index from 8 to 2. - NF removed dissolved organic carbon and the rejection of hardness between 60% and 87%. - The rejection of TDS obtained in the RO system was at a level of 99.7%.	[139]
MF	DCMD	Hot brine City water contaning salt (3.5, 6 or 10%) or Seawater	Organic compounds, Colloids and bacteria	Hydrophobicity of the outside membrane surface was reduced. Very limited flux reduction at salts concentration up to 19.5% from seawater.	[141]
UF	OD	Grape juice	Fermentable sugars and proteins	Increased flux during subsequent concentration of the permeate by OD and wetting reduction because of increase in juice surface tension.	[140]
MF followed by degassing	LGMD	Polluted seawater	Salts, oil, silt, sludge and unknown organic compounds	Distillate with high quality and salt separation factors well above 10,000 have been obtained.	[142]
pH contol using HCl at pH = 4.1	DCMD	Tap water, CaCO$_3$ and mixed CaCO$_3$/CaSO$_4$ solutions	Hardness CaCO$_3$/CaSO$_4$	Quite stable vapor flux during the experiment time (7 h).	[35]
Lime precipitation by Ca(OH)$_2$, sedimentation and filtration	DCMD	Wastewater	CaSO$_4$ and silicon compound	The fouling was significantly diminished.	[143]
Coagulation/flocculation and MF	DCMD	OMW	Solids and organic compounds (Phenolic compounds, Sugar and Proteins)	MF pre-treatment improved significantly the DCMD permeate flux compared to coagulation/flocculation pre-treatment.	[85]
Thermal softening followed by filtration	DCMD	Tap/ground/lake water	Hardness (bicarbonate)	HCO$_3$ ions was lowered 2–3 times by keeping water at the boiling point for 15 min.	[144]
Sedimentation and UF Boiling and MF	DCMD	Bilge water/saline wastewater	Hardness, organic compounds and proteins	Sedimentation and UF showed a significant flux decline. But boiling and MF pretreatment avoided the rapid flux decline.	[33]
Coagulation, filtration, acidification and degasification	DCMD	Recirculating Cooling Water	organic matter (NOM), total phosphorus (TP) and suspended substance (SS)	About 23% of improvement of flux was obtained by using coagulation pretreatment after 30 days of operation.	[145]
UF with coagulation	VMD	RO-concentrated wastewater from steel plant	Mainly hardness and Organic matter	The flux was 30% higher when the pretreatment was used and the CODcr removal reached 40%.	[56]

Recently, Jansen, et al. [142] used filtration (i.e., 10 μm pore size) followed by degassing as pretreatments to LGMD pilot plant and obtained high quality distillate. It was reported that the effect of degassing on the MD process was not conclusive, even though all the results showed that degassing had a positive effect on the permeate flux.

Chemical pretreatment has been focused on bacteria, hardness scale and oxidizing agents for their inhibition, destruction or reduction. It involves several methods such as coagulation, flocculation, precipitation, softening, pH control, etc.

He, et al. [35] applied acidification with HCl (pH = 4.1) in order to mitigate membrane scaling in DCMD. The results showed a quite stable permeate flux during the DCMD operating time (i.e., 7 h) proving that the acidification with HCl reduced the concentration of HCO_3^- or CO_3^{2-} in the feed aqueous solution due to the neutralization by H^+ following this reaction ($H^+ + HCO_3^- \rightarrow H_2O + CO_2$).

El-Abbassi, et al. [85] employed coagulation/flocculation and MF as pre-treatment processes for the treatment of OMW by DCMD. It was found that MF was the optimum pretreatment to be integrated to DCMD for OMW compared to coagulation/flocculation. MF allowed a reduction of about 30% of the total solids, whereas coagulation/flocculation permitted only 23% of the total solids.

Thermal pretreatment was considered to remove most bicarbonate from water, which in turn reduced the amount of precipitate formed during MD process. Thermal softening pretreatment or boiling followed by filtration was used to reduce the concentration of HCO_3^- ion below the level of 0.6 mmol/L (i.e., lowered 2–3 times) by keeping water at the boiling [144]. Gryta [33] investigated the effect of boiling and filtration of saline wastewater containing proteins to limit the intensity of fouling in DCMD. Two pretreatments were investigated: (i) sedimentation followed by UF and the obtained UF permeate was used as a feed for DCMD process, so that the results showed a significant permeate flux decrease; (ii) wastewater was boiled for 30 min and after cooling at room temperature after 12 h it was filtered through a filter paper and the result showed that TOC concentration decreased from 2780 to 1120 mg TOC/dm^3 while the solution turbidity was reduced from 68.1 to 14.3 NTU. This proved that the thermal pretreatment followed by filtration permitted to avoid the rapid permeate flux decline and allowed to remove proteins foulants from the feed, therefore minimizing their precipitation on the membrane surface.

Wang, et al. [145] studied the performance and the effect of coagulation pretreatment on the efficiency of MD process for desalination of recirculating cooling water (RCW). Coagulation, filtration, acidification and degasification units were used in this study. Poly-aluminum chloride (PACl) used as coagulant was effective to remove most of the natural organic matter (NOM), antiscalant additives total phosphorus (TP) substances and suspended substance (SS) in RCW. After the coagulation and sedimentation processes, the RCW was filtered through 5 μm filter and then an acidification treatment was carried out using 0.2 mol/L HCl followed by degassing in order to reduce the CO_2 concentration. The results indicated that when coagulation pretreatment was not employed for desalination by MD process, a rapid decline of the MD permeate flux was observed. However, about 23% improvement of the permeate flux was obtained by using coagulation pretreatment after 30 days operation.

The efficiency of any pretreatment depends on many parameters such as the nature of feed solutions, their foulant(s) properties and the characteristics of the membrane. All pretreatments represent an additional economic cost since either another membrane-based filtration system must be set up or a higher energy cost (i.e., boiling temperature, thermal softening, cooling recirculation, etc.) must be added. However, the importance of a pretreatment in MD cannot be overlooked because it acts as the first strategy plan for membrane fouling prevention.

Independently of the selected pretreatment, the performance of MD and OD processes is always improved when a pretreatment was carried out. However, other actions focused on the feed solution treatment (i.e., use of antiscalants) and on the membrane should be also considered to reduce or prevent fouling phenomena.

5.2. Use of Antiscalants

Antiscalants or scale inhibitors are surface active materials that inhibit inorganic scaling not only at low temperature but also at high temperature processes. These agents can prevent scaling not only in MD but also in OD or OMD [46] specially in pilot-scale desalination or RO brine treatment. The antiscalants are used to minimize the potential of scale formation on membrane surface. Antiscalants interfere with precipitation reactions in three primary ways: threshold inhibition, crystal formation and dispersion [146]. The first one refers to the capacity of an antiscalant to maintain supersaturated solutions of sparingly soluble salts (e.g., $CaCO_3$, $CaSO_4$, $MgCO_3$, $Fe(OH)_3$, $CaF2$, $Fe(OH)_3$, $Al(OH)_3$, etc.). The second one concerns the process of an antiscalants to deform crystal shapes, resulting in soft non adherent scales. The third one means the ability of some antiscalants to adsorb on crystals or colloidal particles and impart a highly negative charge to the crystal thereby keeping them separated and preventing propagation.

Early antiscalants used sodium hexametaphosphate (SHMP) as a threshold agent to inhibit the growth of calcium carbonate and sulfate-based scales. Care must be taken to avoid hydrolysis of SHMP in the dosing tank that may decrease the scale inhibition efficiency. The following reaction can describe the steps of polyphosphate hydrolysis to ortho-phosphate ($PO_3^- + H_2O \rightarrow H_2PO_4^- \rightarrow HPO_4^{2-} \rightarrow PO_4^{3-}$) [146,147]. In addition to the hydrolysis problem there is a calcium phosphate scaling risk that can be described as ($3Ca^{2+} + 2PO_4^{3-} \rightarrow Ca_3(PO_4)_2$).

The most used antiscalants in desalination are derived from three chemical groups [148]. Those including condensed polyphosphates, organophosphonates and polyelectrolytes. Effective polyelectrolyte inhibitors are mostly polycarboxylic acids (e.g., polyacrylic acid, polymethacrylic acid and polymaleic acid) [46]. In general, the most effective antiscalant contains a blend of polyacrylic acid (PAA) and phosphoric acid or polyacrylate and a hydroxyethylidene diphosphonate (HEDP) [46].

Gryta [46] investigated the effect of polyphosphates antiscalant on the formation of $CaCO_3$ using a PP capillary module under DCMD configuration. When the feed without antiscalant was used the results of XRD analysis of the deposit formed on the PP membrane surface showed the calcite polymorphic form of $CaCO_3$ scale. However, when polyphosphates were applied as antiscalant under various compositions (2–20 ppm), practically no $CaCO_3$ crystal was detected on the membrane surface.

According to He, et al. [149] five different commercial antiscalants, namely K797 (water acrylic terpolymer), K752 (polyacrylic acid and sodium polyacrylate based compound), GHR (solution of a nitrogen containing organo-phosphorus compound), GLF (organo-phosphorus antiscalant) and GSI (based on neutralized carboxylic and phosphonic acids), were tested in DCMD process with different concentrations in the range 0.6 to 70 mg/L using a PP hollow fiber membrane coated with fluorosilicone on its external surface. The antiscalant K752 was found to be more effective in inhibiting $CaSO_4$ scaling compared with the other tested antiscalants due to its excellent thermal stability compared with polyphosphates and other copolymers antiscalants. In addition, GHR reduced calcite scaling, whereas not much difference was observed between the performances of the other antiscalants. It is necessary to point out that no wetting problem was detected because the surface tension of antiscalants solutions (71.5 mN/m) was near to that of tap water (71.8 mN/m).

It must be mentioned that many antiscalants molecules, typically amphiphilic molecules, such as surfactants and other amphiphilic organics, often reduce the surface tension of water solutions resulting in membrane pore wetting problems and shortening the life-time of the membrane [92,150]. Therefore, prior to use, it is necessary to determine the surface tension of the feed solutions treated with antiscalants and measure the *LEP* of the membrane using these feed solutions. In addition, it is to be noted that there is a risk of microbiological contamination of antiscalant solutions because some antiscalants are nutrients for microbes, algae and bacteria (e.g., SHMP, orthophosphate). Moreover, other antiscalants containing phosphorous can accelerate the growth of microbes [151].

5.3. Membrane Modification

As previously mentioned, another methodology to prevent membrane fouling is to design fouling resistant membranes. This can be achieved using specific material and membrane surface modification. For instance, hydrophobicity, roughness and charge of the membrane surface are strongly related to fouling due to several interactions or adsorption/desorption between the membrane and the foulant(s). Various surface modification methods such as photochemical land redox grafting, immobilization of nanoparticles, plasma treatment, physical coating with polymers and chemical reactions on the membrane surface have been adopted in order to reduce or inhibit membrane fouling. In general, for MD, OD and OMD, membrane modification should focus on the improvement of both hydrophobicity, membrane surface omniphobicity and *LEP* characteristics to prevent pore wetting [24,27,28].

In order to prepare a superhydrophobic membrane, Razmjou, et al. [152] modified a microporous PVDF membrane by depositing TiO_2 nanoparticles on the membrane surface using a low temperature hydrothermal. Subsequently, the TiO_2 coated membrane was fluorosilanized by H,1H,2H,2H-perfluorododecyltrichlorosilane. The modified membrane showed a good hydrophobicity exhibiting a water contact angle up to 166° compared to new membrane, 125°. Moreover, the *LEP* was increased from 120 kPa to 190 kPa after membrane modification. Fouling mitigation was examined in a DCMD process using HA and NaCl as a feed solution. A 20 h fouling experiment with HA did not show any reduction of the permeate flux for virgin and modified membranes but when 3.8 mM $CaCl_2$ was added in the feed solution a significant permeate flux reduction was observed due to the formation of the complexes with HA. Nevertheless, the modified membrane was less prone to fouling than the unmodified one.

Xu, et al. [153] coated microporous PTFE membranes with sodium alginate hydrogel for OD of oily feeds. The results showed that the transmembrane mass transfer coefficient decreased by less than 5% because of the membrane surface coating and the OD permeate flux, when using 0.2, 0.4 and 0.8 wt % orange oil water mixtures as feed over a period of 300 min, indicated that the coated membranes were resistant to wetting. However, the uncoated membrane was immediately wet by 0.2 wt % orange oil water feed solution.

In another study, Zuo and Wang [154] developed an effective method to modify hydrophobic PVDF membrane to enhance anti-oil fouling property for MD applications. The PVDF flat sheet membrane surface was successfully grafted by polyethylene glycol (PEG) and TiO_2 particles via plasma treatment. Compared to the modified PTFE membranes by Xu, et al. [153], the MD experiments showed that the modified membrane presented a stable water permeate flux over 24 h of operation without oil fouling nor wetting.

Zhang, et al. [155] fabricated a superhydrophobic membrane by spraying a mixture of polydimethylsiloxane (PDMS) and hydrophobic SiO_2 nanoparticles on PVDF flat sheet membranes surfaces. The water contact angle and *LEP* increased from 107° and 210 kPa to 156° and 275 kPa, respectively. The results of the DCMD experiments, which lasted 180 h using 25 wt % NaCl as a feed solution, indicated for the modified membrane a rejection factor above 99.99% with a slight decline of the permeate flux. However, the permeate flux of the unmodified membrane was decreased considerably. In addition, the SEM image indicated good fouling resistance of the modified membrane. Comparatively, An, et al. [30] showed that the hydrophobic PDMS microspheres coated on an electrospun PVDF-HFP membrane improved anti-wetting (i.e., a significant increase of the water contact angle 155.4°) and antifouling properties of the membrane when treating dyes aqueous solutions. This membrane exhibited a more negative charge than that of a commercial PVDF and therefore less fouling to differently-charged dyes.

He, et al. [34], who coated PP hollow fiber membranes with a porous fluorosilicone for their use in DCMD process, claimed that the fluorosilicone coating could discourage surface nucleation and particles attachment. In another study [35], mentioned that the porous fluorosilicone coating layer on the PP hollow fiber membranes was helpful to eliminate and develop the necessary resistance against the deposit scale such as $CaSO_4$ and $CaCO_3$ in DCMD experiments.

It is worth noting that modification of hydrophobic membranes for the improvement of the membrane fouling resistance is less studied compared to that of hydrophilic membranes. In general, there is a lack of information about the effect of the hydrophobic polymer type on the prevention of fouling/scaling in MD processes. The fouling mechanism in MD must be investigated in depth because a systematic study on the interactions of membranes with different feed solutions has not been performed yet.

5.4. Membrane Cleaning

Unlike the use of pretreatments and antiscalants that reduce to the maximum the fouling phenomena, cleaning of fouled membranes is other of the explored strategies to extend the membrane life. Generally, after rinsing the membrane with distilled water, the used chemical cleaning agents include acids, alkalis, surfactants, enzymes and metal chelating agents (i.e., organic compounds that form complexes with metal ions) [15,17,156]. Research studies used different strong and weak acids to clean scales such as HCl, which is particularly effective to remove basic crystal salts (e.g., $CaCO_3$) by dissolving the deposit from the membrane surface [14,49,69]. According to Gryta [49], rinsing the membrane module by 3 wt % HCl allowed to dissolve the $CaCO_3$ scale on the PP hollow fiber membrane and restore the initial membrane permeability. In another study, Gryta [69] used 2–5 wt % HCl solution to rinse the membrane module every 40–80 h of DCMD operation. As a result, the formed deposit, which was mainly $CaCO_3$, was removed and the initial efficiency of the membrane module was recovered. Similarly, Curcio, et al. [157] used two steps to clean a fouled membrane with HA and $CaCO_3$ by using citric acid aqueous solution at pH 4 for 20 min followed by 0.1 M NaOH aqueous solution for 20 min. This cleaning procedure also allowed the complete recovery of the permeate flux and the hydrophobicity of the membrane.

The high efficiency of basic cleaning procedures against the deposit of HA due to its good dissociation and dissolution at high pH values was also confirmed by Srisurichan, et al. [42] who reported that rinsing with distilled water the fouled membrane by HA containing $CaCl_2$, for 2 h followed by 20 min recirculation of 0.1 M NaOH aqueous solution resulted in 100% permeate flux recovery. Guillen-Burrieza, et al. [111] studied membrane fouling and cleaning of solar MD plant following different cleaning strategies to remove the fouling layer and restore the membrane properties (i.e., distilled water at pH = 6.15, 5 wt % sulfuric acid at pH = 1, 5 wt % citric acid at pH = 1.77, 5 wt % formic acid at pH = 1.72, 0.1 wt % $Na_5P_3O_{10}$ + 0.2 wt % EDTA at pH = 6.67 and 0.1 wt % oxalic acid + 0.8 wt % citric acid at pH = 2.2). The use of 0.1 wt % oxalic acid and 0.8 wt % citric acid solution was found to be the most suitable cleaning protocol as it was able to remove a great part of the scaling layer, formed mainly by NaCl, Fe, Mg and Al oxides. After the second cleaning procedure, the distillate quality was improved (i.e., salt rejection factor was 85%).

Durham and Nguyen [82] developed an effective cleaning procedure for PTFE membranes fouled after 2 h of OD processing with 21.5% of tomato puree. The permeate side of the membrane was flushed with fresh water, whereas the feed side of the membrane was rinsed with water at 40–50 °C for 10 min and then the cleaning agents were circulated at the same temperature range for 60 min. The cleaning agents included water, NaOH, HNO_3 and enzymes (lipolase, alcalase, palatase and pectinex) with different concentrations. The enzyme/surfactant cleaning was performed using P3 Ultrasil 25, P3 Ultrasil 30, P3 Ultrasil 53, P3 Ultrasil 56, P3 Ultrasil 60A and 1% P3 Ultrasil 75. It was observed that acidic and enzymatic cleaning agents were ineffective cleaners. However, 1% NaOH was the most effective cleaner for membranes with a surface tension greater than 23 mN/m and 1% P3 Ultrasil56 was the best cleaner too for membranes with a surface tension less than 23 mN/m as it was able to maintain the water permeate flux in addition to the hydrophobicity of the membrane.

Zarebska, et al. [103] investigated the effect of consecutive cleaning of MD membranes used for the treatment of model manure solution by using distilled water, alkaline/acid and Novadan cleaning agents. The results showed that cleaning with distilled water had the lowest cleaning efficiency, whereas cleaning with distilled water followed by either NaOH/citric acid or Novadan agents was more efficient. Among the tested cleaning strategies, it was claimed that Novadan agent was the most successful in removing proteins and carbohydrates from PTFE membrane while it removed only proteins from PP membrane [103].

For cleaning biofouled membranes, the use of biocides [15] followed by rinsing with distilled water can be effective to recover the initial permeate flux. Krivorot, et al. [158] used NaOH at a pH value of about 12 and a temperature of 40 °C to hydrolyze organic/bacterial fouling, followed by distilled water and 70% ethanol for sterilizing the system, and finally distilled water to remove ethanol. It was observed that the initial permeability was recovered by removing the biofilm deposit.

The use of the appropriate cleaning agent depends on the fouling and scaling nature, deposit location and the membrane resistance to chemical cleaning [2,111]. Typical MD polymers (PP, PTFE, PVDF) offer high resistance to chemical cleaning. However, it must be noted that cleaning porous and hydrophobic membranes used in MD, OD and OMD is not an easy task because of the high risk of pore wetting. In fact, generally in MD process, cleaning of membranes was found to be insufficient provided that fouling is also associated with membrane wettability. For instance, pressure and temperature play an important role in membrane cleaning [156]. High temperatures improve cleaning by increasing transport and solubility of the fouling material as well as the reaction kinetics. Minimum pressure cannot force the fouling layer onto the membrane surface, making it less adhesive.

6. Conclusions and Future Perspectives

Membrane fouling is a common problem and complex phenomena in all processes used for seawater desalination and waste water treatment applications. For MD, OD and OMD, fouling leads to membrane pore wetting and blocking. Compared to other membrane separation processes such as the pressure-driven membrane processes (MF, UF, RO, etc.), very few research studies have been published so far on fouling mechanisms in MD, OD and OMD, and investigations on the kinetics behind fouling phenomena and fouling mitigation remain very scarce and poorly understood. Moreover, when fouling is studied, the considered characterization techniques focused only on the average physico-chemical properties of the surface deposits but not on the underlying deposit layers.

In MD, OD and OMD, fouling is influenced by various parameters such as the membrane characteristics, especially the pore size and the material of the membrane surface, operation conditions and nature of the fee aqueous solutions.

As it is well known, MD, OD and OMD membrane technologies suffer from the temperature and concentration polarization phenomena and various strategies have been adopted in order to reduce their effects including the increase of the flow rate of both the feed and permeate solutions, turbulent promoters, etc. These polarization phenomena can have a major influence on inorganic fouling. For salts like $CaCO_3$ and $CaSO_4$ whose solubility in water decreases with increasing temperature, the temperature polarization phenomenon encourages the crystals formation of these salts in the bulk feed solution. However, for salts like NaCl whose solubility in water increases with increasing temperature, the concentration and temperature polarization phenomena encourage the crystals formation of these salts on the membrane surface where the temperature is lower and the concentration is higher.

Greater hydrophobicity and lower pore size tend to increase organic fouling effects. This type of fouling also depends on the nature of the organic matter, the MD operating conditions (temperature, transmembrane pressure, flow rate) and the characteristics of the feed solution (pH, ionic strength).

Biofouling refers to the growth of bacteria or micro-organisms on the membrane surface and to biological particles that may be trapped at both the membrane surface and/or pores forming a biofilm. This type of fouling is the least studied in MD, OD and OMD compared to the other types of fouling.

Fouling investigation in OD has received less attention compared to MD and fouling in OMD may take place following the same mechanisms as those occurred in DCMD and OD fouling, provided that OMD combines both MD and OD.

Many research studies using different membrane characterization techniques still need to be done in order to understand the formation mechanisms of the different fouling types. Some characterization techniques reported in the present review paper (TEM, UTDR, EEM and FFF) are yet to be applied to analyze fouled membranes used in MD, OD and OMD processes. No analytical technique can be used on its own to characterize the fouling deposit on the membrane surface. A combination of different techniques seems to be more appropriate.

Feed pretreatment, membrane modification, use of antiscalants and cleaning strategies of membrane surfaces are the most used methods to diminish or prevent foulant(s) deposition onto the membrane surface and in its pores during MD, OD or OMD applications. Up to now, pretreatment has been the appropriate method to prevent and minimize fouling in MD, OD and OMD processes. Antiscalants inhibit the inorganic fouling (i.e., scaling) and minimize the potential for forming scale layer on the membrane surface. Condensed polyphosphates, organophosphonates and polyelectrolytes are the common antiscalants used in desalination [140]. Attention should be paid to some antiscalants that are nutrients for microbes, algae and bacteria increasing therefore the risk of microbiological contamination.

One explored strategy to extend the membrane life is the chemical cleaning. In general, alkaline agents are advisable for cleaning organic fouled membranes. However, soluble salts like calcium carbonate or iron oxides, require acid cleaners. In addition, membrane surface modification seems to be one of the promising methods for developing anti-fouling membranes as it can enhance both hydrophobicity and *LEP* characteristics of the membranes in order to prevent pore wetting. Generally, in MD, OD and OMD processes, cleaning membranes was found to be insufficient provided that fouling is also associated with membrane wettability.

Although the fouling topic has generated much interest among researchers, this phenomenon still needs to be deeply studied using different characterization techniques applied not only on the membrane surface but also inside its pores. This will permit us to understand this phenomenon well and propose new methods to prevent it. On the other hand, there is a need to design novel and advanced membranes for MD, OD, OMD resistant to fouling.

Author Contributions: The author Mourad Laqbaqbi collected, classified and analysed all data from the cited literature. He wrote the first draft of the paper and followed its revision. All authors contributed in the writing, discussion, analysis and completion of the manuscript.

Conflicts of Interest: The authors declare no conflicts of interest.

References

1. Alkhudhiri, A.; Darwish, N.; Hilal, N. Membrane distillation: A comprehensive review. *Desalination* **2012**, *287*, 2–18. [CrossRef]
2. Khayet, M.; Matsuura, T. *Membrane Distillation: Principles and Aplication*; Elsevier: Amsterdam, The Netherlands, 2011.
3. Chew, J.W.; Krantz, W.B.; Fane, A.G. Effect of a macromolecular- or bio-fouling layer on membrane distillation. *J. Membr. Sci.* **2014**, *456*, 66–76. [CrossRef]
4. Bui, A.V.; Nguyen, H.M.; Muller, J. A laboratory study on glucose concentration by OD in hollow fibre module. *J. Food Eng.* **2004**, *63*, 237–245. [CrossRef]
5. Zambra, C.; Romero, J.; Pino, L.; Saavedra, A.; Sanchez, J. Concentration of cranberry juice by osmotic distillation process. *J. Food Eng.* **2015**, *144*, 58–65. [CrossRef]
6. Cassano, A.; Drioli, E. Concentration of clarified kiwifruit juice by osmotic distillation. *J. Food Eng.* **2007**, *79*, 1397–1404. [CrossRef]
7. Gryta, M. Osmotic MD and other membrane distillation variants. *J. Membr. Sci.* **2005**, *246*, 145–156. [CrossRef]
8. Wang, L.; Min, J. Modeling and analyses of membrane osmotic distillation using non-equilibrium thermodynamics. *J. Membr. Sci.* **2011**, *378*, 462–470. [CrossRef]
9. Ravindra Babu, B.; Rastogi, N.K.; Raghavarao, K.S.M.S. Concentration and temperature polarization effects during osmotic membrane distillation. *J. Membr. Sci.* **2008**, *322*, 146–153. [CrossRef]

10. Raghavarao, K.S.M.S.; Madhusudhan, M.C.; Tavanandi, T.H.; Niranjan, K. *Athermal Membrane Processes for the Concentration of Liquid Foods and Natural Colors. Chapter 12 BOOK Da-Wen Sun-Emerging Technologies for Food Processing*, 2nd ed.; Academic Press, Elsevier: New York, NY, USA, 2014.

11. Khayet, M.; Mengual, J.I. Effect of salt concentration during the treatment of humic acid solutions by membrane distillation. *Desalination* **2004**, *168*, 373–381. [CrossRef]

12. Hausmann, A.; Sanciolo, P.; Vasiljevic, T.; Weeks, M.; Schroën, K.; Gray, S.; Duke, M. Fouling mechanisms of dairy streams during membrane distillation. *J. Membr. Sci.* **2013**, *441*, 102–111. [CrossRef]

13. Nguyen, Q.M.; Lee, S. Fouling analysis and control in a DCMD process for SWRO brine. *Desalination* **2015**, *367*, 21–27. [CrossRef]

14. Warsinger, D.M.; Swaminathan, J.; Guillen-Burrieza, E.; Arafat, H.A.; Lienhard V, J.H. Scaling and fouling in membrane distillation for desalination applications: A review. *Desalination* **2015**, *356*, 294–313. [CrossRef]

15. Tijing, L.D.; Woo, Y.C.; Choi, J.S.; Lee, S.; Kim, S.H.; Shon, H.K. Fouling and its control in membrane distillation—A review. *J. Membr. Sci.* **2015**, *475*, 215–244. [CrossRef]

16. Alklaibi, A.M.; Lior, N. Membrane-distillation desalination: Status and potential. *Desalination* **2005**, *171*, 111–131. [CrossRef]

17. Lee, H.; Amy, G.; Cho, J.; Yoon, Y.; Moon, S.H.; Kim, I.S. Cleaning strategies for flux recovery of an ultrafiltration membrane fouled by natural organic matter. *Water Res.* **2001**, *35*, 3301–3308. [CrossRef]

18. Wang, P.; Chung, T.S. Recent advances in membrane distillation processes: Membrane development, configuration design and application exploring. *J. Membr. Sci.* **2015**, *474*, 39–56. [CrossRef]

19. Khayet, M. Membranes and theoretical modeling of membrane distillation, a review. *Adv. Colloid Interface Sci.* **2011**, *164*, 56–88. [CrossRef] [PubMed]

20. Nane, S.; Patil, G.; Raghavarao, K.S.M.S. Chapter 19: Membrane Distillation in Food Processing. In *Handbook of Membrane Separations: Chemical, Pharmaceutical, Food, and Biotechnological Applications*; Pabby, A.K., Rizvi, S.S.H., Sastre, A.M., Eds.; CRC Press: Boca Raton, FL, USA, 2008.

21. Lawson, K.W.; Lloyd, D.R. Membrane distillation: A review. *J. Membr. Sci.* **1997**, *124*, 1–25. [CrossRef]

22. Drioli, E.; Ali, A.; Macedonio, F. Membrane distillation: Recent developments and perspectives. *Desalination* **2015**, *356*, 56–84. [CrossRef]

23. Eykens, L.; De Sitter, K.; Dotremont, C.; Pinoy, L.; Van der Bruggen, B. How to Optimize the Membrane Properties for Membrane Distillation: A Review. *Ind. Eng. Chem. Res.* **2016**, *55*, 9333–9343. [CrossRef]

24. Hilal, N.; Khayet, M.; Wright, C.J. *Membrane Modification: Technology and Application*; CRC Press, Taylor & Francis Group: Boca Raton, FL, USA, 2012.

25. Mansouri, J.; Fane, A.G. Membrane development for processing of oily feeds in IMD, osmotic distillation: Developments in technology and modelling. In Proceedings of the Workshop on "Membrane Distillation, Osmotic Distillation and Membrane Contactors" (CNRIRMERC), Cetraro, Italy, 2–4 July 1998; pp. 43–46.

26. Cheng, D.Y.; Wiersma, S.J. Composite Membrane for Membrane Distillation System. U.S. Patent 4,419,242, 23 February 1982.

27. Boo, C.; Lee, J.; Elimelech, M. Engineering Surface Energy and Nanostructure of Microporous Films for Expanded Membrane Distillation Applications. *Environ. Sci. Technol.* **2016**, *50*, 8112–8119. [CrossRef] [PubMed]

28. Lee, J.; Boo, C.; Ryu, W.H.; Taylor, A.D.; Elimelech, M. Development of Omniphobic Desalination Membranes Using a Charged Electrospun Nanofiber Scaffold. *ACS Appl. Mater. Interfaces* **2016**, *8*, 11154–11161. [CrossRef] [PubMed]

29. An, A.K.; Guo, J.; Jeong, S.; Lee, E.J.; Tabatabai, S.A.; Leiknes, T. High flux and antifouling properties of negatively charged membrane for dyeing wastewater treatment by membrane distillation. *Water Res.* **2016**, *103*, 362–371. [CrossRef] [PubMed]

30. An, A.K.; Guo, J.; Lee, E.J.; Jeong, S.; Zhao, Y.; Wang, Z.; Leiknes, T. PDMS/PVDF hybrid electrospun membrane with superhydrophobic property and drop impact dynamics for dyeing wastewater treatment using membrane distillation. *J. Membr. Sci.* **2017**, *525*, 57–67. [CrossRef]

31. Khayet, M.; Velázquez, A.; Mengual, J.I. Direct contact membrane distillation of humic acid solutions. *J. Membr. Sci.* **2004**, *240*, 123–128. [CrossRef]

32. Narayan, A.V.; Nagaraj, N.; Hebbar, H.U.; Chakkaravarthi, A.; Raghavarao, K.S.M.S.; Nene, S. Acoustic field-assisted osmotic membrane distillation. *Desalination* **2002**, *147*, 149–156. [CrossRef]

33. Gryta, M. Fouling in direct contact membrane distillation process. *J. Membr. Sci.* **2008**, *325*, 383–394. [CrossRef]

34. He, F.; Gilron, J.; Lee, S.; Sirkar, K.K. Potential for scaling by sparingly soluble salts in cross flow DCMD. *J. Membr. Sci.* **2008**, *311*, 68–80. [CrossRef]

35. He, F.; Sirkar, K.K.; Gilron, J. Studies on scaling of membranes in desalination by direct contact membrane distillation: CaCO₃ and mixed CaCO₃/CaSO₄ systems. *Chem. Eng. Sci.* **2009**, *64*, 1844–1859. [CrossRef]

36. Jacob, P.; Phungsai, P.; Fukushi, K.; Visvanathan, C. Direct contact membrane distillation for anaerobic effluent treatment. *J. Membr. Sci.* **2015**, *475*, 330–339. [CrossRef]

37. Hausmann, A.; Sanciolo, P.; Vasiljevic, T.; Weeks, M.; Schroën, K.; Gray, S.; Duke, M. Fouling of dairy components on hydrophobic polytetrafluoroethylene (PTFE) membranes for membrane distillation. *J. Membr. Sci.* **2013**, *442*, 149–159. [CrossRef]

38. Ding, Z.; Liu, L.; Liu, Z.; Ma, R. Fouling resistance in concentrating TCM extract by direct contact membrane distillation. *J. Membr. Sci.* **2010**, *362*, 317–325. [CrossRef]

39. Naidu, G.; Jeong, S.; Vigneswaran, S. Interaction of humic substances on fouling in membrane distillation for seawater desalination. *Chem. Eng. J.* **2015**, *262*, 946–957. [CrossRef]

40. Naidu, G.; Jeong, S.; Kim, S.J.; Kim, I.S.; Vigneswaran, S. Organic fouling behavior in direct contact membrane distillation. *Desalination* **2014**, *347*, 230–239. [CrossRef]

41. Guillen-Burrieza, E.; Thomas, R.; Mansoor, B.; Johnson, D.; Hilal, N.; Arafat, H. Effect of dry-out on the fouling of PVDF and PTFE membranes under conditions simulating intermittent seawater membrane distillation (SWMD). *J. Membr. Sci.* **2013**, *438*, 126–139. [CrossRef]

42. Srisurichan, S.; Jiraratananon, R.; Fane, A.G. Humic acid fouling in the membrane distillation process. *Desalination* **2005**, *174*, 63–72. [CrossRef]

43. Ge, J.; Peng, Y.; Li, Z.; Chen, P.; Wang, S. Membrane fouling and wetting in a DCMD process for RO brine concentration. *Desalination* **2014**, *344*, 97–107. [CrossRef]

44. Mokhtar, N.M.; Lau, W.J.; Ismail, A.F.; Veerasamy, D. Membrane Distillation Technology for Treatment of Wastewater from Rubber Industry in Malaysia. *Procedia CIRP* **2015**, *26*, 792–796. [CrossRef]

45. Yu, X.; Yang, H.; Lei, H.; Shapiro, A. Experimental evaluationon concentrating cooling tower blow down water by direct contact membrane distillation. *Desalination* **2013**, *323*, 134–141. [CrossRef]

46. Gryta, M. Polyphosphates used for membrane scaling inhibition during water desalination by membrane distillation. *Desalination* **2012**, *285*, 170–176. [CrossRef]

47. Gryta, M. Influence of polypropylene membrane surface porosity on the performance of membrane distillation process. *J. Membr. Sci.* **2007**, *287*, 67–78. [CrossRef]

48. Gryta, M.; Tomaszewska, M.; Grzechulska, J.; Morawski, A.W. Membrane distillation of NaCl solution containing natural organic matter. *J. Membr. Sci.* **2001**, *181*, 279–287. [CrossRef]

49. Gryta, M. Alkaline scaling in the membrane distillation process. *Desalination* **2008**, *228*, 128–134. [CrossRef]

50. Gryta, M.; Grzechulska-Damszel, J.; Markowska, A.; Karakulski, K. The influence of polypropylene degradation on the membrane wettability during membrane distillation. *J. Membr. Sci.* **2009**, *326*, 493–502. [CrossRef]

51. Tian, R.; Gao, H.; Yang, X.H.; Yan, S.Y.; Li, S. A new enhancement technique on air gap membrane distillation. *Desalination* **2014**, *332*, 52–59. [CrossRef]

52. Alkhudhiri, A.; Darwish, N.; Hilal, N. Treatment of high salinity solutions: Application of air gap membrane distillation. *Desalination* **2012**, *287*, 55–60. [CrossRef]

53. Mericq, J.P.; Laborie, S.; Cabassud, C. Vacuum membrane distillation of seawater reverse osmosis brines. *Water Res.* **2010**, *44*, 5260–5273. [CrossRef] [PubMed]

54. Zhao, Z.P.; Zhu, C.Y.; Liu, D.Z.; Liu, W.F. Concentration of ginseng extracts aqueous solution by vacuum membrane distillation 2. Theory analysis of critical operating conditions and experimental confirmation. *Desalination* **2011**, *267*, 147–153. [CrossRef]

55. Ji, Z.; Wang, J.; Hou, D.; Yin, Z.; Luan, Z. Effect of microwave irradiation on vacuum membrane distillation. *J. Membr. Sci.* **2013**, *429*, 473–479. [CrossRef]

56. Zhiqing, Y.; Xiaolong, L.; Chunrui, W.; Xuan, W. Effect of pretreatment on membrane fouling and VMD performance in the treatment of RO- concentrated wastewater. *Desalin. Water Treat.* **2013**, *51*, 6994–7003. [CrossRef]

57. Banat, F.; Al-Asheh, S.; Qtaishat, M. Treatment of waters colored with methylene blue dye by vacuum membrane distillation. *Desalination* **2005**, *174*, 87–96. [CrossRef]

58. Zarebska, A.; Nieto, D.R.; Christensen, K.V.; Norddahl, B. Ammonia recovery from agricultural wastes by membrane distillation: Fouling characterization and mechanism. *Water Res.* **2014**, *56*, 1–10. [CrossRef] [PubMed]

59. Criscuoli, A.; Zhong, J.; Figoli, A.; Carnevale, M.C.; Huang, R.; Drioli, E. Treatment of dye solutions by vacuum membrane distillation. *Water Res.* **2008**, *42*, 5031–5037. [CrossRef] [PubMed]

60. Yuan, W.; Zydney, A.L. Humic acid fouling during microfiltration. *J. Membr. Sci.* **1999**, *157*, 1–12. [CrossRef]

61. Jermann, D.; Pronk, W.; Meylan, S.; Boller, M. Interplay of different NOM fouling mechanisms during ultrafiltration for drinking water production. *Water Res.* **2007**, *41*, 1713–1722. [CrossRef] [PubMed]

62. Nilson, J.A.; Digiano, F.A. Influence of NOM composition on nanofiltration. *J. Am. Water Works Assoc.* **1996**, *88*, 53–66.

63. Hong, S.; Elimelech, M. Chemical and physical aspects of natural organic matter (NOM) fouling of nanofiltration membranes. *J. Membr. Sci.* **1997**, *132*, 159–181. [CrossRef]

64. Gryta, M. Concentration of saline wastewater from the production of heparin. *Desalination* **2000**, *129*, 35–44. [CrossRef]

65. Hamrouni, B.; Dhahbi, M. Calco-carbonic equilibrium calculation. *Desalination* **2002**, *152*, 167–174. [CrossRef]

66. Wirth, D.; Cabassud, C. Water desalination using membrane distillation: Comparison between inside/out and outside/in permeation. *Desalination* **2002**, *147*, 139–145. [CrossRef]

67. Safavi, M.; Mohammadi, T. High-salinity water desalination using VMD. *Chem. Eng. J.* **2009**, *149*, 191–195. [CrossRef]

68. Gryta, M.; Tomaszewska, M.; Karakulski, K. Wastewater treatment by membrane distillation. *Desalination* **2006**, *198*, 67–73. [CrossRef]

69. Gryta, M. Long-term performance of membrane distillation process. *J. Membr. Sci.* **2005**, *265*, 153–159. [CrossRef]

70. Tun, C.M.; Fane, A.G.; Matheickal, J.T.; Sheikholeslami, R. Membranes distillation crystallization of concentrated salts-flux and crystal formation. *J. Membr. Sci.* **2005**, *257*, 144–155. [CrossRef]

71. Yun, Y.; Ma, R.; Zhang, W.; Fane, A.G.; Li, J. Direct contact membrane distillation mechanism for high concentration NaCl solutions. *Desalination* **2006**, *188*, 251–262. [CrossRef]

72. Mariah, L.; Buckley, C.A.; Brouckaert, C.J.; Curcio, E.; Drioli, E.; Jaganyi, D.; Ramjugernath, D. Membrane distillation of concentrated brines-Role of water activities in the evaluation of driving force. *J. Membr. Sci.* **2006**, *280*, 937–947. [CrossRef]

73. Drioli, E.; Criscuoli, A.; Curcio, E. Integrated membrane operation for seawater desalination. *Desalination* **2002**, *147*, 77–81. [CrossRef]

74. Curcio, E.; Profio, G.D.; Drioli, E. Membrane crystallization of macromolecular solutions. *Desalination* **2002**, *145*, 173–177. [CrossRef]

75. Bouguecha, S.; Dhabi, M. Fluidized bed crystallizer and air gap membrane distillation as a solution to geothermal water desalination. *Desalination* **2002**, *152*, 237–244. [CrossRef]

76. Gryta, M. The assessment of microorganism growth in the membrane distillation system. *Desalination* **2002**, *142*, 79–88. [CrossRef]

77. Vogt, M.; Flemming, H.; Veeman, W. Diffusion in Pseudomonas aeruginosa biofilms: A pulsed field gradient NMR study. *J. Biotechnol.* **2000**, *77*, 137–146. [CrossRef]

78. Flemming, H.C.; Schaule, G.; McDonogh, R.; Ridgway, H.F. Effects and extent of biofilm accumulation in membrane systems. In *Biofouling and Biocorrosion in Industrial Water Systems*; Geesey, G.G., Lewandowsky, Z., Flemming, H.C., Eds.; CRC Press/Lewis Publishers: Boca Raton, FL, USA, 1994; pp. 63–89.

79. Goh, S.; Zhang, J.; Liu, Y.; Fane, A.G. Fouling and wetting in membrane distillation (MD) and MD-bioreactor (MDBR) for wastewater reclamation. *Desalination* **2013**, *323*, 39–47. [CrossRef]

80. Goh, S.; Zhang, Q.; Zhang, J.; McDougald, D.; Krantz, W.B.; Liu, Y.; Fane, A.G. Impact of a biofouling layer on the vapor pressure driving force and performance of a membrane distillation process. *J. Membr. Sci.* **2013**, *438*, 140–152. [CrossRef]

81. El-Abbassi, A.; Khayet, M.; Kiai, H.; Hafidi, A.; García-Payo, M.C. Treatment of crude olive mill wastewaters by osmotic distillation and osmotic membrane distillation. *Sep. Purif. Technol.* **2013**, *104*, 327–332. [CrossRef]

82. Durham, R.J.; Nguyen, H.M. Hydrophobic membrane evaluation and cleaning for osmotic distillation of tomato puree. *J. Membr. Sci.* **1994**, *87*, 181–189. [CrossRef]
83. Kujawski, W.; Sobolewska, A.; Jarzynka, K.; Güell, C.; Ferrando, M.; Warczok, J. Application of osmotic membrane distillation process in red grape juice concentration. *J. Food Eng.* **2013**, *116*, 801–808. [CrossRef]
84. Bui, A.V.; Nguyen, H.M. Scaling up of osmotic distillation from laboratory to pilot plant for concentration of fruit juices. *Int. J. Food Eng.* **2005**, *1*, 1556–3758. [CrossRef]
85. El-Abbassi, A.; Hafidi, A.; Khayet, M.; García-Payo, M.C. Integrated direct contact membrane distillation for olive mill wastewater treatment. *Desalination* **2013**, *323*, 31–38. [CrossRef]
86. Ruiz Salmón, I.; Janssens, R.; Luis, P. Mass and heat transfer study in osmotic membrane distillation-crystallization for CO2 valorization as sodium carbonate. *Sep. Purif. Technol.* **2017**, *176*, 173–183. [CrossRef]
87. Bui, A.V.; Nguyen, H.M.; Joachim, M. Characterisation of the polarisations in osmotic distillation of glucose solutions in hollow fibre module. *J. Food Eng.* **2005**, *68*, 391–402. [CrossRef]
88. Karlsson, E.; Luh, B.S. Vegetable juices, sauces and soups. In *Commercial Vegetable Processing*; Luh, B.S., Woodroof, J.G., Eds.; Medtech: New York, NY, USA, 1988.
89. Wills, R.B.H.; Lim, J.S.K.; Greenfield, H. Composition of Australian foods. *Tomato Food Thechnol. Aust.* **1984**, *36*, 78–80.
90. Charfi, A.; Jang, H.; Kim, J. Membrane fouling by sodium alginate in high salinity conditions to simulate biofouling during seawater desalination. *Bioresour. Technol.* **2017**, in press. [CrossRef] [PubMed]
91. Lokare, O.R.; Tavakkoli, S.; Wadekar, S.; Khanna, V.; Vidic, R.D. Fouling in direct contact membrane distillation of produced water from unconventional gas extraction. *J. Membr. Sci.* **2017**, *524*, 493–501. [CrossRef]
92. Wang, Z.; Lin, S. Membrane fouling and wetting in membrane distillation and their mitigation by novel membranes with special wettability. *Water Res.* **2017**, *112*, 38–47. [CrossRef] [PubMed]
93. Zhao, F.; Chu, H.; Yu, Z.; Jiang, S.; Zhao, X.; Zhou, X.; Zhang, Y. The filtration and fouling performance of membranes with different pore sizes in algae harvesting. *Sci. Total Environ.* **2017**. [CrossRef] [PubMed]
94. Girão, A.V.; Caputo, G.; Ferro, M.C. Application of Scanning Electron Microscopy–Energy Dispersive X-Ray Spectroscopy (SEM-EDS). *Compr. Anal. Chem.* **2017**, *75*. [CrossRef]
95. Jafarzadeh, Y.; Yegani, R.; Sedaghat, M. Preparation, characterization and fouling analysis of ZnO/polyethylene hybrid membranes for collagen separation. *Chem. Eng. Res. Des.* **2015**, *94*, 417–427. [CrossRef]
96. Shirazi, S.; Lin, C.J.; Chen, D. Inorganic fouling of pressure-driven membrane processes—A critical review. *Desalination* **2010**, *250*, 236–248. [CrossRef]
97. Liu, Z.; Ohsuna, T.; Sato, K.; Mizuno, T.; Kyotani, T.; Nakane, T.; Terasaki, O. Transmission electron microscopy observation on fine structure of zeolite NaA membrane. *Chem. Mater.* **2006**, *18*, 922–927. [CrossRef]
98. Heinzl, C.; Ossiander, T.; Gleich, S.; Scheu, C. Transmission electron microscopy study of silica reinforced polybenzimidazole membranes. *J. Membr. Sci.* **2015**, *478*, 65–74. [CrossRef]
99. Taheri, M.L.; Stach, E.A.; Arslan, I.; Crozier, P.A.; Kabius, B.C.; LaGrange, T.; Minor, A.M.; Takeda, S.; Tanase, M.; Wagner, J.B.; et al. Current status and future directions for in situ transmission electron microscopy. *Ultramicroscopy* **2016**, *170*, 86–95. [CrossRef] [PubMed]
100. Weiss, C.; McLoughlin, P.; Cathcart, H. Characterisation of dry powder inhaler formulations using atomic force microscopy. A Review. *Int. J. Pharm.* **2015**, *494*, 393–407. [CrossRef] [PubMed]
101. Khulbe, K.C.; Feng, C.Y.; Matsuura, T. *Synthetic Polymeric Membranes: Characterization by Atomic Force Microscopy*; Springer: Heidelberg, Germany, 2008.
102. Bowen, W.R.; Hilal, N. *Atomic Force Microscopy in Process Engineering: An Introduction to AFM for Improved Processes and Products*; Elsevier: Oxford, UK, 2009.
103. Zarebska, A.; Amor, Á.C.; Ciurkot, K.; Karring, H.; Thygesen, O.; Andersen, T.P.; Hägg, M.-B.; Christensen, K.V.; Norddahl, B. Fouling mitigation in membrane distillation processes during ammonia stripping from pig manure. *J. Membr. Sci.* **2015**, *484*, 119–132. [CrossRef]
104. Yun, M.A.; Yeon, K.M.; Park, J.S.; Lee, C.H.; Chun, J.; Lim, D.J. Characterization of biofilm structure and its effect on membrane permeability in MBR for dye wastewater treatment. *Water Res.* **2006**, *40*, 45–52. [CrossRef] [PubMed]

105. Canette, A.; Briandet, R. Confocal Laser Scanning Microscopy. *Agro ParisTech* **2014**, *2*, 1389–1396.
106. Yuan, B.; Wang, X.; Tang, C.; Li, X.; Yu, G. In situ observation of the growth of biofouling layer in osmotic membrane bioreactors by multiple fluorescence labeling and confocal laser scanning microscopy. *Water Res.* **2015**, *75*, 188–200. [CrossRef] [PubMed]
107. Ferrando, M.; Rrzek, A.; Zator, M.; Lopez, F.; Guell, C. An approach to membrane fouling characterization by confocal scanning laser microscopy. *J. Membr. Sci.* **2005**, *250*, 283–293. [CrossRef]
108. West, S.; Horn, H.; Hijnen, W.A.M.; Castillo, C.; Wagner, M. Confocal laser scanning microscopy as a tool to validate the efficiency of membrane cleaning procedures to remove biofilms. *Sep. Purif. Technol.* **2014**, *122*, 402–411. [CrossRef]
109. Spettmann, D.; Eppmann, S.; Flemming, H.-C.; Wingender, J. Visualization of membrane cleaning using confocal laser scanning microscopy. *Desalination* **2008**, *224*, 195–200. [CrossRef]
110. Courel, M.; Tronel-Peyroz, E.; Rios, G.M.; Dornier, M.; Reynes, M. The problem of membrane characterization for the process of osmotic distillation. *Desalination* **2001**, *140*, 15–25. [CrossRef]
111. Guillen-Burrieza, E.; Ruiz-Aguirre, A.; Zaragoza, G.; Arafat, H.A. Membrane fouling and cleaning in long term plant-scale membrane distillation operations. *J. Membr. Sci.* **2014**, *468*, 360–372. [CrossRef]
112. Sanmartino, J.A.; Khayet, M.; García-Payo, M.C.; El Bakouri, H.; Riaza, A. Desalination and concentration of saline aqueous solutions up to supersaturation by air gap membrane distillation and crystallization fouling. *Desalination* **2016**, *393*, 39–51. [CrossRef]
113. Bagavathiappan, S.; Lahiri, B.B.; Saravanan, T.; Philip, J.; Jayakumar, T. Infrared thermography for condition monitoring—A review. *Infrared Phys. Technol.* **2013**, *60*, 35–55. [CrossRef]
114. Marinetti, S.; Cesaratto, P.G. Emissivity estimation for accurate quantitative thermography. *NDT & E Int.* **2012**, *51*, 127–134.
115. Ndukaife, K.O.; Ndukaife, J.C.; Agwu Nnanna, A.G. Membrane fouling characterization by infrared thermography. *Infrared Phys. Technol.* **2015**, *68*, 186–192. [CrossRef]
116. Li, X.; Zhang, H.; Hou, Y.; Gao, Y.; Li, J.; Guo, W.; Ngo, H.H. In situ investigation of combined organic and colloidal fouling for nanofiltration membrane using ultrasonic time domain reflectometry. *Desalination* **2015**, *362*, 43–51. [CrossRef]
117. Mairal, A.P.; Greenberg, A.R.; Krantz, W.B.; Bond, L.J. Real time measurement of inorganic fouling of RO desalination membranes using ultrasonic time domain reflectometry. *J. Membr. Sci.* **1999**, *159*, 185–196. [CrossRef]
118. Zhang, Z.X.; Greenberg, A.R.; Krantz, W.B.; Chai, G.Y. Study of membrane fouling and cleaning in spiral wound modules using ultrasonic time-domain reflectometry. In *New Insights into Membrane Science and Technology Polymeric, Inorganic and Biofunctional Membranes*; Butterfield, A.A., Bhattacharyya, D., Eds.; Elsevier: Amsterdam, The Netherlands, 2003; pp. 65–88.
119. Taheri, A.H.; Sim, S.T.V.; Sim, L.N.; Chong, T.H.; Krantz, W.B.; Fane, A.G. Development of a new technique to predict reverse osmosis fouling. *J. Membr. Sci.* **2013**, *448*, 12–22. [CrossRef]
120. Xu, X.; Li, J.; Li, H.; Cai, Y.; Cao, Y.; He, B.; Zhang, Y. Non-invasive monitoring of fouling in hollow fiber membrane via UTDR. *J. Membr. Sci.* **2009**, *326*, 103–110. [CrossRef]
121. Tung, K.-L.; Teoh, H.-C.; Lee, C.-W.; Chen, C.-H.; Li, Y.-L.; Lin, Y.-F.; Chen, C.-L.; Huang, M.-S. Characterization of membrane fouling distribution in a spiral wound module using high-frequency ultrasound image analysis. *J. Membr. Sci.* **2015**, *495*, 489–501. [CrossRef]
122. Hunter, R.J. *Zeta Potential in Colloid Science: Principles and Applications*; Academic Press: London, UK, 1981.
123. Mikhaylin, S.; Bazinet, L. Fouling on ion-exchange membranes: Classification, characterization and strategies of prevention and control. A review. *Adv. Colloid Interface Sci.* **2016**, *229*, 34–56. [CrossRef] [PubMed]
124. Erickson, D.; Li, D. Streaming Potential and Streaming Current Methods for Characterizing Heterogeneous Solid Surfaces. *J. Colloid Interface Sci.* **2001**, *237*, 283–289. [CrossRef] [PubMed]
125. Tanaka, K.; Kodama, S.; Goto, T. *X-ray Diffraction Studies on the Deformation and Fracture of Solids*; Elsevier: Amsterdam, The Netherlands, 1993; Volume 10.
126. Kim, J.; Kwon, H.; Lee, S.; Lee, S.; Hong, S. Membrane distillation (MD) integrated with crystallization (MDC) for shale gas produced water (SGPW) treatment. *Desalination* **2017**, *403*, 172–178. [CrossRef]
127. Melián-Martel, N.; Sadhwani, J.J.; Malamis, S.; Ochsenkühn-Petropoulou, M. Structural and chemical characterization of long-term reverse osmosis membrane fouling in a full scale desalination plant. *Desalination* **2012**, *305*, 44–53. [CrossRef]

128. Beckhoff, B.; Kanngießer, B.; Langhoff, N.; Wedell, R.; Wolff, H. *Handbook of Practical X-ray Fluorescence Analysis*; Springer: Berlin/Heidelberg, Germany, 2006.
129. Thygesen, O.; Hedegaard, M.A.B.; Zarebska, A.; Beleites, C.; Krafft, C. Membrane fouling from ammonia recovery analyzed by ATR-FTIR imaging. *Vib. Spectrosc.* **2014**, *72*, 119–123. [CrossRef]
130. Tomaszewska, M.; Białończyk, L. Influence of proteins content in the feed on the course of membrane distillation. *Desalin. Water Treat.* **2013**, *51*, 2362–2367. [CrossRef]
131. Nguyen, Q.-M.; Jeong, S.; Lee, S. Characteristics of membrane foulants at different degrees of SWRO brine concentration by membrane distillation. *Desalination* **2017**, *409*, 7–20. [CrossRef]
132. Matthews, B.J.H.; Jones, A.C.; Theodorou, N.K.; Tudhope, A.W. Excitation-emission-matrix fluorescence spectroscopy applied to humic acid bands in coral reefs. *Mar. Chem.* **1996**, *55*, 317–332. [CrossRef]
133. Huang, W.; Chu, H.; Dong, B.; Liu, J. Evaluation of different algogenic organic matters on the fouling of microfiltration membranes. *Desalination* **2014**, *344*, 329–338. [CrossRef]
134. Messaud, F.A.; Sanderson, R.D.; Runyon, J.R.; Otte, T.; Pasch, H.; Williams, S.K.R. An overview on field-flow fractionation techniques and their applications in the separation and characterization of polymers—A review. *Polym. Sci.* **2009**, *34*, 351–368.
135. Giddings, J.C. Field-flow fractionation: Analysis of macromolecular, colloidal, and particulate materials. *Science* **1993**, *260*, 1456–1465. [CrossRef] [PubMed]
136. Lee, E.; Shon, H.K.; Cho, J. Biofouling characteristics using flow field-flow fractionation: Effect of bacteria and membrane properties. *Bioresour. Technol.* **2010**, *101*, 1487–1493. [CrossRef] [PubMed]
137. Karakulski, K.; Gryta, M.; Sasim, M. Production of process water using integrated membrane processes. *Chem. Pap.* **2006**, *60*, 416–421. [CrossRef]
138. Schäfer, A.; Fane, A.; Waite, T. *Nanofiltration: Principles and Applications*; Elsevier: Oxford, UK, 2005.
139. Karakulski, K.; Gryta, M.; Morawski, A. Membrane processes used for potable water quality improvement. *Desalination* **2002**, *145*, 315–319. [CrossRef]
140. Bailey, A.F.G.; Barbe, A.M.; Hogan, P.A.; Johnson, R.A.; Sheng, J. The effect of ultrafiltration on the subsequent concentration of grape juice by osmotic distillation. *J. Membr. Sci.* **2000**, *164*, 195–204. [CrossRef]
141. Song, L.; Ma, Z.; Liao, X.; Kosaraju, P.B.; Irish, J.R.; Sirkar, K.K. Pilot plant studies of novel membrane sand devices for direct contact membrane distillation-based desalination. *J. Membr. Sci.* **2008**, *323*, 257–270. [CrossRef]
142. Jansen, A.E.; Assink, J.W.; Hanemaaijer, J.H.; Medevoort, J.v.; Sonsbeek, E.v. Development and pilot testing of full-scale membrane distillation modules for deployment of waste heat. *Desalination* **2013**, *323*, 55–65. [CrossRef]
143. Gryta, M.; Karakulski, K.; Tomaszewska, M.; Morawski, A. Treatment of effluents from the regeneration of ion exchangers using the MD process. *Desalination* **2005**, *180*, 173–180. [CrossRef]
144. Gryta, M. Desalination of thermally softened water by membrane distillation process. *Desalination* **2010**, *257*, 30–35. [CrossRef]
145. Wang, J.; Qua, D.; Tie, M.; Ren, H.; Peng, X.; Luan, Z. Effect of coagulation pretreatment on membrane distillation process for desalination of recirculating cooling water. *Sep. Purif. Technol.* **2008**, *64*, 108–115. [CrossRef]
146. Kucera, J. *Reverse Osmosis: Design, Processes, and Applications for Engineers*; Wiley: New York, NY, USA, 2010.
147. Prihasto, N.; Liu, Q.; Kim, S. Pre-treatment strategies for seawater desalination by reverse osmosis system. *Desalination* **2009**, *249*, 308–316. [CrossRef]
148. Ketrane, R.; Saidant, R.; Gil, O.; Leleyter, L.; Baraud, F. Efficiency of five scale inhibitors on calciumcarbonate precipitation fromhardwater: Effect of temperature and concentration. *Desalination* **2009**, *249*, 1397–1404. [CrossRef]
149. He, F.; Sirkar, K.K.; Gilron, J. Effects of antiscalants to mitigate membrane scaling by direct contact membrane distillation. *J. Membr. Sci.* **2009**, *345*, 53–58. [CrossRef]
150. Franken, A.C.M.; Nolten, J.A.M.; Mulder, M.H.V.; Bargeman, D.; Smolders, C.A. Wetting criteria for the applicability of membrane distillation. *J. Membr. Sci.* **1987**, *33*, 315–328. [CrossRef]
151. Al-Shammiri, M.; Safar, M.; Al-Dawas, M. Evaluation of Two Different Antisclant in Real Operation at the Doha. *Res. Plant* **2000**, *128*, 1–16.

152. Razmjou, A.; Arifin, E.; Dong, G.; Mansouri, J.; Chen, V. Superhydrophobic modification of TiO$_2$ nanocomposite PVDF membranes for applications in membrane distillation. *J. Membr. Sci.* **2012**, *415–416*, 850–863. [CrossRef]

153. Xu, J.B.; Lange, S.; Bartley, J.P.; Johnson, R.A. Alginate-coated microporous PTFE membranes for use in the osmotic distillation of oily feeds. *J. Membr. Sci.* **2004**, *240*, 81–89. [CrossRef]

154. Zuo, G.; Wang, R. Novel membrane surface modification to enhance anti-oil fouling property for membrane distillation application. *J. Membr. Sci.* **2013**, *447*, 26–35. [CrossRef]

155. Zhang, J.; Song, Z.; Li, B.; Wang, Q.; Wang, S. Fabrication and characterization of superhydrophobic poly (vinylidene fluoride) membrane for direct contact membrane distillation. *Desalination* **2013**, *324*, 1–9. [CrossRef]

156. Al-Amoudi, A.; Lovitt, R.W. Fouling strategies and the cleaning system of NF membranes and factors affecting cleaning efficiency. *J. Membr. Sci.* **2007**, *303*, 4–28. [CrossRef]

157. Curcio, E.; Ji, X.; Di Profio, G.; Sulaiman, A.O.; Fontananova, E.; Drioli, E. Membrane distillation operated at high seawater concentration factors: Role of the membrane on CaCO$_3$ scaling in presence of humic acid. *J. Membr. Sci.* **2010**, *346*, 263–269. [CrossRef]

158. Krivorot, M.; Kushmaro, A.; Oren, Y.; Gilron, J. Factors affecting biofilm formation and biofouling in membrane distillation of seawater. *J. Membr. Sci.* **2011**, *376*, 15–24. [CrossRef]

applied
sciences

MDPI

Article

Hydrophobic Ceramic Membranes for Water Desalination

Joanna Kujawa [1], Sophie Cerneaux [2], Wojciech Kujawski [1,*] and Katarzyna Knozowska [1]

[1] Faculty of Chemistry, Nicolaus Copernicus University in Torun, 7 Gagarina St., 87-100 Torun, Poland; joanna.kujawa@umk.pl (J.K.); kasiaknozowska@wp.pl (K.K.)

[2] Institut Europeen des Membranes, UMR 5635, Place Eugene Bataillon, 34095 Montpellier CEDEX 5, France; Sophie.Cerneaux@univ-montp2.fr

* Correspondence: wojciech.kujawski@umk.pl; Tel.: +48-56-611-43-15; Fax: +48-56-611-45-26

Academic Editor: Enrico Drioli

Received: 26 January 2017; Accepted: 13 April 2017; Published: 15 April 2017

Abstract: Hydrophilic ceramic membranes (tubular and planar) made of TiO_2 and Al_2O_3 were efficiently modified with non-fluorinated hydrophobic grafting molecules. As a result of condensation reaction between hydroxyl groups on the membrane and reactive groups of modifiers, the hydrophobic surfaces were obtained. Ceramic materials were chemically modified using three various non-fluorinated grafting agents. In the present work, the influence of grafting time and type of grafting molecule on the modification efficiency was evaluated. The changes of physicochemical properties of obtained hydrophobic surfaces were determined by measuring the contact angle (CA), roughness (RMS), and surface free energy (SFE). The modified surfaces were characterized by contact angle in the range of 111–132°. Moreover, hydrophobic tubular membranes were utilized in air-gap membrane distillation to desalination of sodium chloride aqueous solutions. The observed permeate fluxes were in the range of 0.7–4.8 $kg \cdot m^{-2} \cdot h^{-1}$ for tests with pure water. The values of permeate fluxes for membranes in contact with NaCl solutions were smaller, within the range of 0.4–2.8 $kg \cdot m^{-2} \cdot h^{-1}$. The retention of NaCl in AGMD process using hydrophobized ceramic membranes was close to unity for all investigated membranes.

Keywords: ceramic membranes; non-fluorinated alkylsilanes; air-gap membrane distillation

1. Introduction

Membrane distillation (MD) is an emerging non-isothermal membrane separation process, being one of the promising techniques for the desalination of highly saline waters [1–7]. Contrary to pressure-driven techniques, the driving force in the membrane distillation process is related to the difference in chemical potential (generated by difference of temperatures) between the two sides of the hydrophobic membrane [8–10]. The non-wetted porous hydrophobic membranes are required for an efficient process realization. There are various MD modes utilized, for example, sweep gas membrane distillation (SGMD), direct contact membrane distillation (DCMD), air gap membrane distillation (AGMD), vacuum membrane distillation (VMD) [1,5,9,11]. MD can be used in different applications such as waste water treatment, desalination, and food processing [1,3,5,11–19]. In all configurations, the liquid–vapor equilibrium is the determining factor yielding to the selectivity of the MD processes. The mass transfer in MD follows three subsequent steps: liquid–vapor phase transition at the membrane pores entrance on feed side; transfer of vapors through the pores of the membrane; and condensation of vapors on the permeate side of the membrane [1–7].

Despite the fact that, in the MD process, high pressure is not required and the process offers a very promising performance for both the stand-alone and the desalination process, full-scale commercialization of MD still copes with various problems. These difficulties are related to the lack of suitable and effective module design, proper membranes, as well as intensive energy consumption (in cases when waste heat, solar, or different alternative energy sources are not utilized). In the presented article, the authors address the issues related to the lack of suitable membranes. The major requirements for the membrane features are related to hydrophobic character and porous structure [20]. During the MD process only vapors of solvent (water) are allowed to pass across the membrane. Concerning scientific literature focused on MD application, it is possible to find application of commercially-available hollow fiber or flat-sheet microfiltration membranes [16,21–29]. The utilization of these membranes in MD is associated with their hydrophobic properties, decent porosity, as well as adequate pore sizes. Nevertheless, these membranes are not the perfectly-designed and dedicated to membrane distillation, therefore they suffer from wetting problems as well as low permeability and short life span [9,21–26,30].

To design and form hydrophobic surfaces with certain properties, two approaches can be selected. One possibility is the creation of a rough surface with a structure responsible for its hydrophobicity. By generating heterogeneity on the surface, it is possible to form highly hydrophobic surfaces due to generation of pillars and the presence of air pockets on the surface. The other way is utilization of chemical modification applying grafting agent materials possessing the low surface free energy [31,32]. Up till now, a number of methods have been developed to produce rough surfaces, like solidification of melted alkylketene dimer (AKD) [33], plasma polymerization/etching of polypropylene in the presence of polytetrafluoroethylene (PTFE) [34], microwave plasma-enhanced chemical vapor deposition (MWPE-CVD) of trimethylmethoxysilane (TMMOS) [35], anodic oxidization of aluminum [36], or an immersion of porous alumina gel films into boiling water [37]. Chemically attached coupling agents with low-surface-energy can be also efficiently used to turn the hydrophilic character materials to hydrophobic one [38–41]. Within that group of compounds, perfluoroalkylsilane (PFAS) molecules [42–45], hydrophobic polymers [46], as well as Grignard compounds [47,48] have been found to generate hydrophobic surfaces. The aforementioned grafting agents were efficiently modified by the chemical attachment to the ceramic support. Ceramic materials are characterized by high chemical, thermal, and mechanical stability; therefore, they are ideal materials for many applications in the chemical, biotechnological [49,50], food [51], and pharmaceutical industries [52,53], as well as in water and wastewater processing [40,41,49]. However, ceramic membranes are hydrophilic by nature. This material property limits the wider application of ceramic membranes. For that particular reason, the modification of membrane surface is required to change the surface character from hydrophilic to hydrophobic one. Considering the high effectiveness of the grafting process of ceramic materials and good performance in membrane-based separation techniques (e.g., vacuum pervaporation [42,43,54,55], vacuum membrane distillation [56], air-gap and direct membrane distillation [42,45], and organic solvent nanofiltration [47,48,57]) it should be pointed out that such modification also possessed disadvantages. Namely, a significant drawback is the presence of fluorine atoms in coupling agents. In the presented research fluorine-free modifiers were used for highly efficient hydrophobization process of ceramic membranes. Subsequently, the ceramic materials are tested in desalination process using membrane distillation (MD) technique. An important issue of the study was to evaluate the influence of the type of grafting agents and modification conditions on the membrane performance in MD.

2. Materials and Methods

2.1. Materials

Al_2O_3 (Pall Exekia, Bazet, France) and TiO_2 (TAMI Industries, Nyons, France) tubular ceramic membranes were used in the presented work. Alumina membranes possessed 100 nm pore size (MWCO $\approx$ 150 kDa), whereas titania ones were characterized by 100 kD ($\approx$75 nm) molecular weight cut-off (MWCO). Both types of membranes were characterized by 10/5 mm of outer/inner diameters and 150 mm length. Additionally, TiO$_2$ 100 kD planar membranes (TAMI Industries, France) were used for material characterization.

The following grafting compounds were purchased from Linegal Chemicals (Poland): *n*-octyltrichlorosilane (C8Cl3); *n*-octyltriethoxysilane (C8OEt3); and trichloro(octadecyl)silane (C18Cl3) were used. Chloroform (stabilized by 1% ethanol), acetone, ethanol, *n*-butanol, *n*-hexane, and glycerin were purchased from Avantor Performance Materials (Poland). All compounds and chemicals were used as received.

2.2. Experimental Protocol of Ceramic Membranes Modification

Grafting solutions (0.05 M) were prepared by dissolving an appropriate amount of alkylsilanes in chloroform (stabilized by 1% ethanol). The preparation steps and modification were performed under inert gas (argon) atmosphere to avoid polycondensation reaction [31–41]. Prior to grafting, the ceramic membranes (tubular and/or planar) were cleaned consecutively in ethanol, acetone, and distilled water for 10 min in each solvent and dried in an oven at 110 °C for 12 h. Subsequently, ceramic supports were modified by soaking samples in the grafting solution for a given period of time equal to 1.5 h followed by the second modification of 3 h what resulted in the total hydrophobization time equal to 4.5 h. Scheme of modification procedure is depicted in Figure 1. Subsequently, the grafting effectiveness was evaluated by the contact angle measurements in the case of planar membranes and by determination of water liquid entry pressure (LEPw) for tubular ceramics.

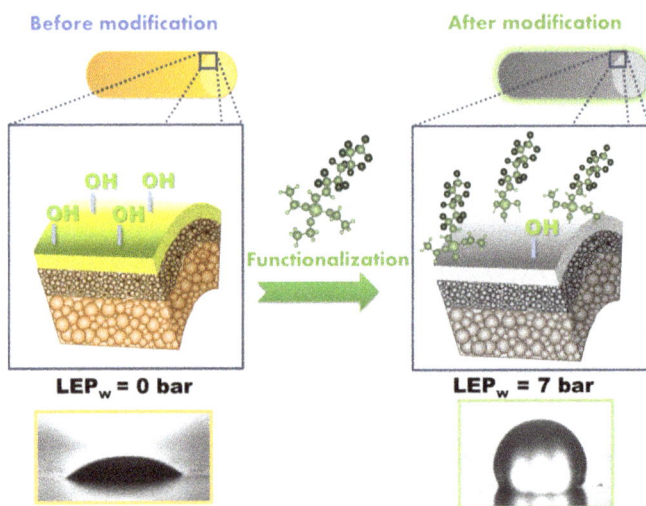

Figure 1. Ceramic membrane modification by alkylsilanes grafting agent and differences in hydrophobicity level (contact angle and water liquid entry pressure (LEPw)) before (left) and after (right) hydrophobization process.

2.3. Modified Membrane Characterization—Analytical Methods

Contact Angle (CA) and Surface Free Energy (SFE): Static and dynamic contact angles (CA) measurements were performed at room temperature using the goniometer PG-X (FibroSystem AB) and deionized water (18 MΩ cm) and glycerin as testing liquids. CAs were determined for pristine and grafted membranes, based on sessile drop (static CA) and the tilting plate (dynamic CA) methods described elsewhere [44,45]. The apparent CA values were calculated by ImageJ software (ImageJ, NIH—freeware version), with an accuracy of ±2°. Additionally, a contact goniometric analysis was implemented for surface free energy (SFE) assessment for planar ceramic membranes. SFE was calculated based on the Owens–Wendt method [58]. The results are presented as an average value obtained from 20 to 30 measurements (average accuracy of ±5%).

Atomic force microscopy (AFM): This technique was used for surface analysis (topography and phase analysis) of planar membranes (NanoScope MultiMode SPM System and NanoScope IIIa Quadrex controller—Veeco, Digital Instrument, Saint Ives, Cambridgeshire, UK). Tip scanning mode was applied for surface roughness analysis. Ambient temperature conditions were kept during all experiments. The root mean squared (RMS) roughness was used as a parameter describing heterogeneity of the samples. Scan size areas were equal to 5 × 5 μm. All samples were analyzed at least five times and an average value of RMS was calculated (accuracy ±5%).

Liquid entry pressure (LEPw): The grafting effectiveness of tubular membranes was assessed by liquid entry pressure determination (LEPw)—Equation (1). LEPw is a pressure value at which liquid penetrates across open pores of the membrane and is transported through the hydrophobic layer on the permeate side [11,23,42,59]. According to the Laplace–Young equation (Equation (1)), it can be noticed that LEPw value refer to the surface tension (γ_L), contact angle (θ_{ef}) on membrane surface, pore radius (r), and tortuosity factor (B).

$$LEP_w = -B\frac{2\gamma_L}{r}\cos\theta_{ef} \tag{1}$$

The LEPw measurements were realized using a laboratory experimental rig presented in detail elsewhere [42]. LEPw values were determined for all tubular membrane samples, prior to the membrane characterization in membrane distillation process. During the LEP measurements, the time interval between each pressure step was equal to 60 min.

2.4. Air-Gap Membrane Distillation (AGMD)

Membranes efficiency was evaluated in desalination process using air-gap membrane distillation (AGMD) technique and experimental rig presented in Figure 2. The AGMD experimental setup and the detailed experimental protocol are described elsewhere [42,45]. AGMD process was realized at temperate of feed equal to T_f = 90 °C and permeate equal to T_p = 5 °C. The air gap width was equal to 5 mm (Figure 2). Measurements were realized for pure water and for the following feed concentrations of NaCl aqueous solutions: 0.25, 0.5, 0.75, and 1.0 M. Rejection coefficient of sodium chloride (Equation (2)) was controlled by ion chromatography (Dionex DX-100 Ion Chromatograph).

$$R_{NaCl} = \left(1 - \frac{C_p}{C_f}\right) \times 100 \; (\%) \tag{2}$$

where C_p and C_f stand for the NaCl concentration in permeate and feed solution, respectively.

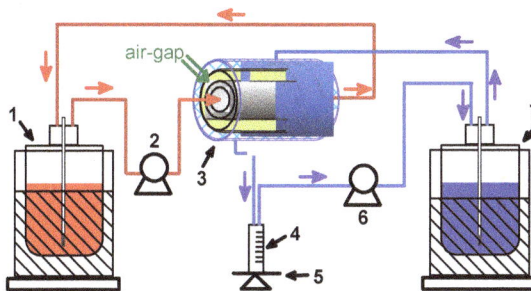

Figure 2. Scheme of a setup used in AGMD experiments (1—thermostated feed tank; 2 and 6—pump; 3—AGMD thermostated membrane module with air gap = 5 mm; 4—measuring cylinder; 5—balance; 7—cooling system).

3. Results and Discussion

3.1. Pristine Ceramic Membranes

Planar unmodified TiO_2 100 kD membrane possessed hydrophilic character with a surface contact angle value equal to $40 \pm 2°$ and total surface free energy equal to 140 ± 5 mN·m^{-1} (polar part 44.8 ± 1.4 mN·m^{-1}; dispersive part 95.2 ± 3.1 mN·m^{-1}). Roughness of unmodified ceramic sample expressed by RMS was equal to 60.5 ± 5.1 nm. Pristine tubular membranes TiO_2 75 nm and Al_2O_3 100 nm were characterized by LEPw equal to 0 bar.

3.2. Effectiveness of Membrane Hydrophobization

The hydrophilic character of planar membrane was turned to hydrophobic one with high efficiency. As a result of hydrophobization, the changes in morphology and physicochemistry of the membrane surface are clearly noticeable (Table 1). All membrane samples possessed contact angle values higher than 90°. The significant increase of water contact angle confirms the formation of the hydrophobic surface. Based on results presented in Table 1, it can be concluded that modification conditions (time and type of grafting agent) have an essential impact on the resulting hydrophobicity level. The highest value of CA was found for samples modified by C18Cl3 molecules, i.e., the molecules with the longest hydrocarbon chain (~2.3 nm). Moreover, comparing ceramics grafted by silanes possessing the same length of alkyl chains, higher CA value was found for C8Cl3 than for C8OEt3. This observation is related to the bond dissociation energy which is equal to 489.5 kJ mol^{-1} and 510.5 kJ mol^{-1} for Si–Cl and Si–OEt, respectively [60]. Similar relation was found for ceramic powders (TiO_2 and ZrO_2) modified by non-fluorinated compounds possessing various reactive groups (e.g., methoxy, ethoxy, and chlorine atoms) [44]. The grafting process affects physicochemistry of ceramics, which was proved by the resulting reduction of surface free energy (SFE) value. It was found that polar component (γ^p) in overall SFE is very small (Table 1). Small contribution of the polar part compared to the dispersive component (γ^d) of SFE is characteristic for hydrophobic and highly hydrophobic materials [42,45,58,61]. Hydrophobization process also influences the surface roughness. The roughness parameter is reduced after grafting and it decreased even more after the extension of grafting duration. Silanization generated smoother surfaces with higher level of hydrophobicity. Namely, the smoothest surface possessing simultaneously the lowest value of overall SFE was achieved for samples modified by C18Cl3 during 4.5 h.

Table 1. Characterization of modified planar ceramic membrane.

Parameter	Pristine	Ti-C8OEt3		Ti-C8Cl3		Ti-C18Cl3	
		1.5 h	4.5 h	1.5h	4.5 h	1.5 h	4.5 h
Contact angle (CA), deg	40 ± 2	111 ± 2	113 ± 2	120 ± 2	128 ± 2	126 ± 2	132 ± 2
SFE, mN·m^{-1}	140 ± 5	62.5 ± 2.7	53.6 ± 2.3	43.7 ± 1.9	32.7 ± 1.4	38.2 ± 1.7	31.1 ± 1.4
Polar part of SFE (γ^p), mN·m^{-1}	44.8 ± 1.4	13.3 ± 0.6	11.4 ± 0.5	9.3 ± 0.4	8.6 ± 0.4	5.7 ± 0.3	5.5 ± 0.2
Dispersive part of SFE (γ^d), mN·m^{-1}	95.2 ± 3.1	49.2 ± 2.2	42.2 ± 1.8	34.4 ± 1.5	24.1 ± 1.1	32.5 ± 1.4	25.6 ± 1.1
RMS, nm	60.5 ± 5.1	50.9 ± 2.0	45.2 ± 1.8	39.6 ± 1.6	33.2 ± 1.3	27.8 ± 1.1	22.9 ± 0.9

The efficiency of tubular membranes modification was assessed by determining LEPw values for alumina and titania membranes after 1.5 h and 4.5 h of modification (Figure 3). A substantial influence of grafting time on the LEPw values, especially for samples grafted by C18Cl3 molecules (Figure 3) was observed. After 1.5 h of hydrophobization LEPw value equal to 1 bar for alumina and titania membranes were observed. Longer exposure of membranes to the grafting agent resulted in the increase of LEPw values to 7 bar and 4 bar for Al_2O_3 and TiO_2 membrane, respectively. The highlighted differences between ceramics can be linked to the membrane material and availability of different amounts of hydroxyl groups available to generate covalent bonds between ceramic surface and grafting coupling agent. Concerning the material properties of ceramics and data presented elsewhere [44], it was found that alumina is richer in hydroxyl groups comparing with titania. For this reason, higher value of LEPw for Al_2O_3 has been noticed. A slightly higher value of LEPw for C8OEt3 than C8Cl3 after a longer modification time might be related to the presence of hydroxyl groups still available after the first grafting process that were not used for covalent bonding.

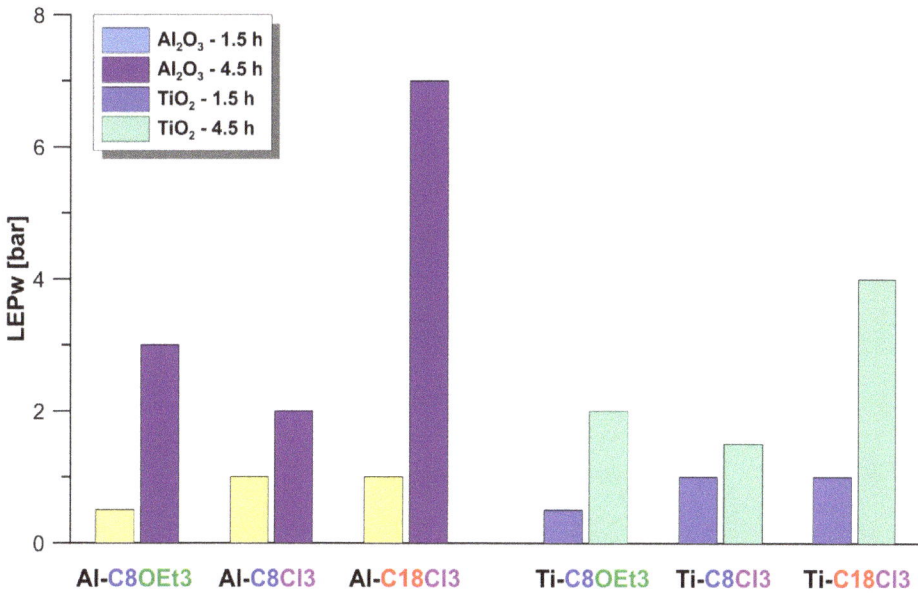

Figure 3. LEPw evolution with grafting time for alumina and titania ceramic materials grafted with various alkylsilanes.

3.3. Air-Gap Membrane Distillation—Separation Efficiency of Modified Ceramic Membranes

Prior to desalination process of NaCl solutions, membranes were tested in contact with pure water as a feed solution to evaluate the nominal value of water permeate flux (Figure 4). The fluxes in contact with pure water were the higher that those for NaCl solutions. Membrane grafted by C8COEt3 possessed the highest value of permeate flux (Figure 4). On the other hand, the less permeable was the membrane grafted by molecules having the longest alkyl chains. The permeate flux through the grafted ceramic membranes decreased with increasing salt concentration in the feed (Figure 4). This behavior was caused by the fact that, in membrane distillation of NaCl solutions, only vapors of water are transported across the hydrophobic porous structure of the membrane and with higher concentration of non-volatile compounds in the feed (sodium chloride) a decline of the flux is observed. For this reason, with increase of salt concertation a diminution of driving can be noticed, according to the Raoult's law.

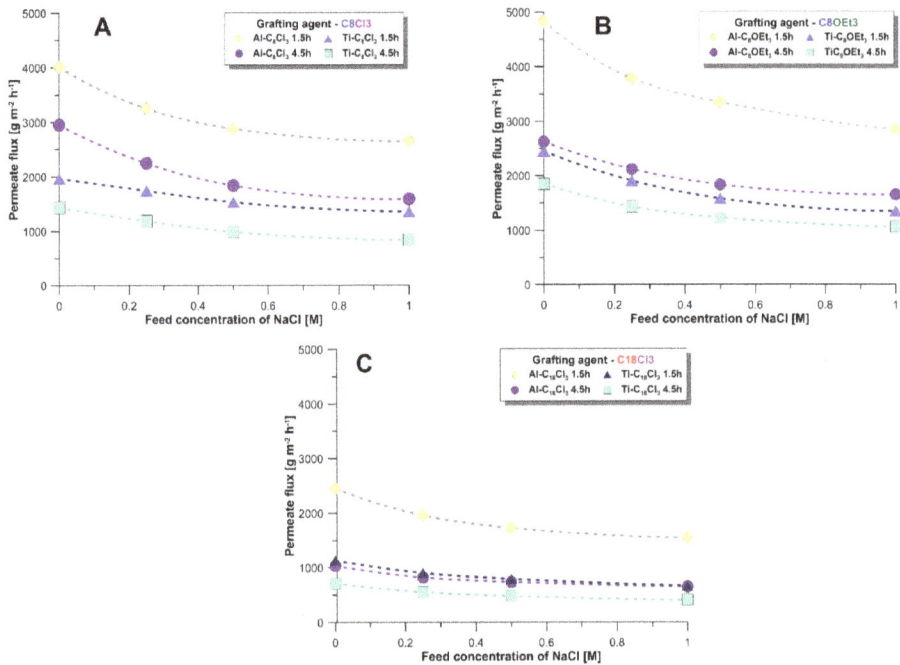

Figure 4. Transport properties of modified ceramic membranes by (**A**) C8Cl3; (**B**) C8OEt3; and (**C**) C18Cl3, molecules in AGMD process used for desalination. T_f = 90 °C; T_p = 5 °C, Δp = 688 mbar.

After a shorter grafting (1.5 h), the alumina membrane modified by C8OEt3 molecules showed the highest value of permeate flux 4.84 kg m^{-2}·h^{-1} which corresponded to the lowest value of measured LEPw (Figure 4B). On the other hand, a membrane sample hydrophobized by molecules with longer alkyl chain (C18Cl) showed the poorest transport properties. This can be attributed to the presence of alkyl chain in a tangled environment and/or partially located in pores [56].

The Al-C18Cl3 membranes produced the lowest value of permeate flux. This observation can be explain by a higher effectiveness of grafting comparing molecules with reactive groups of chlorine atoms and ethoxyl groups. As mentioned above, molecules with chlorine atoms are characterized by the highest ability to generate covalent bonds due to the lower value of energy bonds [60]. The membrane material as well as time of modification process affected the transport properties. Generally, membranes made from alumina possessed a slightly higher value of permeate flux due to slightly bigger pores (100 nm) in comparison with titania ones (≈75 nm). The observed reduction of permeate fluxes (Figure 4) as well as LEPw values (Figure 3) with extended grafting time contributed to formation of a smoother surface (Table 1) showing better water-proof character. Summarizing the fabricated hydrophobic membranes possess good transport properties in AGMD compared with literature data [40,42,62,63]. Furthermore, the observed values of permeate fluxes are higher than for ceramic membranes tested in DCMD. This is related to the benefit of AGMD, namely the reduction of conduction heat losses. In AGMD, permeate is condensed on a chilled surface rather than directly in the chilled permeate. Nevertheless, the achieved permeate fluxes for ceramic membranes are smaller than for polymeric one [64].

Retention coefficient of salt (R_{NaCl}) Equation (2) is a very important factor determining efficiency of membrane distillation process. R_{NaCl} corresponds to the separation properties of applied membranes. The salt retention (R_{NaCl}) during the membrane distillation process was calculated according to Equation (2). The obtained values of R_{NaCl} were very high, close to 100% (Table 2). A small effect of grafting time on the R_{NaCl} values was observed. On the other hand, no influence of membrane materials on their selective properties was noticed. The observed small alteration in R_{NaCl} values can be explained by marginal differences in physiochemistry of the membrane (CA, LEPw, and RMS). Nevertheless, the value of R_{NaCl} was high, proving that no wetting of the membranes was observed.

Table 2. Range of retention coefficient (%) of NaCl in AGMD process.

Membrane	1.5 h Modification	4.5 h Modification
Ti-C8Cl3	98.0–98.3	99.5–99.6
Ti-C8OEt3	98.3–98.5	99.6–99.7
Ti-C18Cl3	98.0–98.4	99.2–99.6
Al-C8Cl3	98.1–98.3	99.0–99.5
Al-C8OEt3	98.0–98.3	99.0–99.3
Al-C18Cl3	98.0–98.2	99.0–99.3

4. Conclusions

The surface character of Al_2O_3 and TiO_2 ceramic membranes (tubular and planar) was changed from a hydrophilic to hydrophobic one by chemical modification with silane grafting agents possessing various structures (length of alkyl chain and type of reactive groups). After modification, highly hydrophobic surfaces were fabricated.

The length of hydrophobic PFAS molecule has a significant impact on the hydrophobicity level. Much higher values of CA and LEPw were observed for planar and tubular membranes grafted by C18 molecules comparing with C8 ones.

Hydrophobization process changed the physicochemical properties of ceramic membranes expressed by their roughness, surface free energy (polar and dispersive component), and hydrophobicity level.

The obtained hydrophobic tubular ceramic membranes were used in membrane separation process—air gap membrane distillation. In AGMD, process impact of concentration of NaCl in the feed, grafting time, as well as type of membrane materials on the transport properties was observed and discussed in detail. Higher values of permeate flux were observed for alumina membranes than for titiania ones, which resulted from higher LEPw values for grafted alumina. The retention coefficient of NaCl in AGMD process using hydrophobized ceramic membranes was over 98% for all investigated membranes.

Acknowledgments: This research was supported by NN 209 255138 grant from the Polish Ministry of Science and Higher Education and by 2012/07/N/ST4/00378 (Preludium 4) grant from the National Science Centre. This research was also partially supported by statutory funds of Nicolaus Copernicus University in Toruń (Faculty of Chemistry, T-109 "Membranes and membrane separation processes—fundamental and applied research").

Author Contributions: Joanna Kujawa and Wojciech Kujawski conceived and designed the experiments; Joanna Kujawa and Sophie Cerneaux performed experiments and collaborated with the data analysis; Katarzyna Knozowska collaborated with interpretation of the experimental data (contact angle, membrane distillation), assisted with critical corrections in the manuscript drafting, and prepared the drawings; Joanna Kujawa and Wojciech Kujawski participated in the analysis and interpretation of the experimental data, cooperated with the drafting and correction of the final version of the manuscript.

Conflicts of Interest: The authors declare no conflict of interest.

References

1. Woo, Y.C.; Tijing, L.D.; Shim, W.-G.; Choi, J.-S.; Kim, S.-H.; He, T.; Drioli, E.; Shon, H.K. Water desalination using graphene-enhanced electrospun nanofiber membrane via air gap membrane distillation. *J. Membr. Sci.* **2016**, *520*, 99–110. [CrossRef]
2. Quist-Jensen, C.A.; Macedonio, F.; Conidi, C.; Cassano, A.; Aljlil, S.; Alharbi, O.A.; Drioli, E. Direct contact membrane distillation for the concentration of clarified orange juice. *J. Food Eng.* **2016**, *187*, 37–43. [CrossRef]

3. Wang, P.; Chung, T.-S. Recent advances in membrane distillation processes: Membrane development, configuration design and application exploring. *J. Membr. Sci.* **2015**, *474*, 39–56. [CrossRef]
4. Drioli, E.; Ali, A.; Simone, S.; Macedonio, F.; Al-Jlil, S.A.; Al Shabonah, F.S.; Al-Romaih, H.S.; Al-Harbi, O.; Figoli, A.; Criscuoli, A. Novel pvdf hollow fiber membranes for vacuum and direct contact membrane distillation applications. *Sep. Purif. Technol.* **2013**, *115*, 27–38. [CrossRef]
5. Quist-Jensen, C.A.; Macedonio, F.; Horbez, D.; Drioli, E. Reclamation of sodium sulfate from industrial wastewater by using membrane distillation and membrane crystallization. *Desalination* **2017**, *401*, 112–119. [CrossRef]
6. El-Bourawi, M.S.; Ding, Z.; Ma, R.; Khayet, M. A framework for better understanding membrane distillation separation process. *J. Membr. Sci.* **2006**, *285*, 4–29. [CrossRef]
7. Alkhudhiri, A.; Darwish, N.; Hilal, N. Membrane distillation: A comprehensive review. *Desalination* **2012**, *287*, 2–18. [CrossRef]
8. Shannon, M.A.; Bohn, P.W.; Elimelech, M.; Georgiadis, J.G.; Marinas, B.J.; Mayes, A.M. Science and technology for water purification in the coming decades. *Nature* **2008**, *452*, 301–310. [CrossRef] [PubMed]
9. Drioli, E.; Wu, Y. Membrane distillation: An experimental study. *Desalination* **1985**, *53*, 339–346. [CrossRef]
10. Drioli, E.; Wu, Y.; Calabro, V. Membrane distillataion in the treatment of aqueous solutions. *J. Membr. Sci.* **1987**, *33*, 277–284. [CrossRef]
11. Zhang, H.; Liu, M.; Sun, D.; Li, B.; Li, P. Evaluation of commercial PTFE membranes for desalination of brine water through vacuum membrane distillation. *Chem. Eng. J.* **2016**, *110*, 52–63. [CrossRef]
12. Tong, D.; Wang, X.; Ali, M.; Lan, C.Q.; Wang, Y.; Drioli, E.; Wang, Z.; Cui, Z. Preparation of Hyflon AD60/PVDF composite hollow fiber membranes for vacuum membrane distillation. *Sep. Purif. Technol.* **2016**, *157*, 1–8. [CrossRef]
13. Rácz, G.; Kerker, S.; Schmitz, O.; Schnabel, B.; Kovács, Z.; Vatai, G.; Ebrahimi, M.; Czermak, P. Experimental determination of liquid entry pressure (LEP) in vacuum membrane distillation for oily wastewaters. *Membr. Water Treat.* **2015**, *6*, 237–249. [CrossRef]
14. Kujawa, J.; Guillen-Burrieza, E.; Arafat, H.A.; Kurzawa, M.; Wolan, A.; Kujawski, W. Raw juice concentration by osmotic membrane distillation process with hydrophobic polymeric membranes. *Food Bioprocess Technol.* **2015**, *8*, 2146–2158. [CrossRef]
15. Guo, F.; Servi, A.; Liu, A.; Gleason, K.K.; Rutledge, G.C. Desalination by membrane distillation using electrospun polyamide fiber membranes with surface fluorination by chemical vapor deposition. *ACS Appl. Mater. Interfaces* **2015**, *7*, 8225–8232. [CrossRef] [PubMed]
16. Criscuoli, A.; Carnevale, M.C. Desalination by vacuum membrane distillation: The role of cleaning on the permeate conductivity. *Desalination* **2015**, *365*, 213–219. [CrossRef]
17. Gryta, M. The application of membrane distillation for broth separation in membrane bioreactors. *J. Membr. Sci. Res.* **2016**, *2*, 193–200.
18. Dong, G.; Kim, J.F.; Kim, J.H.; Drioli, E.; Lee, Y.M. Open-source predictive simulators for scale-up of direct contact membrane distillation modules for seawater desalination. *Desalination* **2017**, *402*, 72–87. [CrossRef]
19. Gryta, M.; Tomaszewska, M.; Karakulski, K. Wastewater treatment by membrane distillation. *Desalination* **2006**, *198*, 67–73. [CrossRef]
20. Eykens, L.; De Sitter, K.; Dotremont, C.; Pinoy, L.; Van der Bruggen, B. How to optimize the membrane properties for membrane distillation: A review. *Ind. Eng. Chem. Res.* **2016**, *55*, 9333–9343. [CrossRef]
21. Peydayesh, M.; Kazemi, P.; Bandegi, A.; Mohammadi, T.; Bakhtiari, O. Treatment of bentazon herbicide solutions by vacuum membrane distillation. *J. Water Process Eng.* **2015**, *8*, e17–e22. [CrossRef]
22. Li, X.; Yu, X.; Cheng, C.; Deng, L.; Wang, M.; Wang, X. Electrospun superhydrophobic organic/inorganic composite nanofibrous membranes for membrane distillation. *ACS Appl. Mater. Interfaces* **2015**, *7*, 21919–21930. [CrossRef] [PubMed]
23. Guillen-Burrieza, E.; Servi, A.; Lalia, B.S.; Arafat, H.A. Membrane structure and surface morphology impact on the wetting of md membranes. *J. Membr. Sci.* **2015**, *483*, 94–103. [CrossRef]
24. Xu, W.-T.; Zhao, Z.-P.; Liu, M.; Chen, K.-C. Morphological and hydrophobic modifications of PVDF flat membrane with silane coupling agent grafting via plasma flow for VMD of ethanol–water mixture. *J. Membr. Sci.* **2015**, *491*, 110–120. [CrossRef]
25. Chen, K.; Xiao, C.; Huang, Q.; Liu, H.; Liu, H.; Wu, Y.; Liu, Z. Study on vacuum membrane distillation (VMD) using fep hollow fiber membrane. *Desalination* **2015**, *375*, 24–32. [CrossRef]

26. Zaragoza, G.; Ruiz-Aguirre, A.; Guillén-Burrieza, E. Efficiency in the use of solar thermal energy of small membrane desalination systems for decentralized water production. *Appl. Energy* **2014**, *130*, 491–499. [CrossRef]

27. Drioli, E.; Ali, A.; Macedonio, F. Membrane distillation: Recent developments and perspectives. *Desalination* **2015**, *356*, 56–84. [CrossRef]

28. Michalkiewicz, B.; Ziebro, J.; Tomaszewska, M. Preliminary investigation of low pressure membrane distillation of methyl bisulphate from its solutions in fuming sulphuric acid combined with hydrolysis to methanol. *J. Membr. Sci.* **2006**, *286*, 223–227. [CrossRef]

29. Orecki, A.; Tomaszewska, M.U.; Morawski, A.W. Treatment of natural waters by nanofiltration. *Przem. Chem.* **2006**, *85*, 1067–1070.

30. Drioli, E. Guest editorial for the special issue: Membrane distillation and related membrane systems. *Desalination* **2013**, *323*, 1. [CrossRef]

31. Ahmad, N.A.; Leo, C.P.; Ahmad, A.L. Superhydrophobic alumina membrane by steam impingement: Minimum resistance in microfiltration. *Sep. Purif. Technol.* **2013**, *107*, 187–194. [CrossRef]

32. Bico, J.; Thiele, U.; Quéré, D. Wetting of textured surfaces. *Colloids Surf. A Physicochem. Eng. Asp.* **2002**, *206*, 41–46. [CrossRef]

33. Onda, T.; Shibuichi, S.; Satoh, N.; Tsujii, K. Super-water-repellent fractal surfaces. *Langmuir* **1996**, *12*, 2125–2127. [CrossRef]

34. Chen, W.; Fadeev, A.Y.; Hsieh, M.C.; Öner, D.; Youngblood, J.; McCarthy, T.J. Ultrahydrophobic and ultralyophobic surfaces: Some comments and examples. *Langmuir* **1999**, *15*, 3395–3399. [CrossRef]

35. Wu, Y.; Sugimura, H.; Inoue, Y.; Takai, O. Thin films with nanotextures for transparent and ultra water-repellent coatings produced from trimethylmethoxysilane by microwave plasma cvd. *Chem. Vap. Depos.* **2002**, *8*, 47–50. [CrossRef]

36. Tsujii, K.; Yamamoto, T.; Onda, T.; Shibuichi, S. Super oil-repellent surfaces. *Angew. Chem. Int. Ed.* **1997**, *36*, 1011–1012. [CrossRef]

37. Tadanaga, K.; Katata, N.; Minami, T. Formation process of super-water-repellent Al_2O_3 coating films with high transparency by the sol–gel method. *J. Am. Ceram. Soc.* **1997**, *80*, 3213–3216. [CrossRef]

38. Alami Younssi, S.; Iraqi, A.; Rafiq, M.; Persin, M.; Larbot, A.; Sarrazin, J. γ alumina membranes grafting by organosilanes and its application to the separation of solvent mixtures by pervaporation. *Sep. Sci. Technol.* **2003**, *32*, 175–179. [CrossRef]

39. Kujawa, J.; Cerneaux, S.; Kujawski, W. Investigation of the stability of metal oxide powders and ceramic membranes grafted by perfluoroalkylsilanes. *Colloids Surf. A Physicochem. Eng. Asp.* **2014**, *443*, 109–117. [CrossRef]

40. Kujawa, J.; Kujawski, W.; Koter, S.; Jarzynka, K.; Rozicka, A.; Bajda, K.; Cerneaux, S.; Persin, M.; Larbot, A. Membrane distillation properties of TiO_2 ceramic membranes modified by perfluoroalkylsilanes. *Desalin. Water Treat.* **2013**, *51*, 1352–1361. [CrossRef]

41. Kujawa, J.; Kujawski, W.; Koter, S.; Rozicka, A.; Cerneaux, S.; Persin, M.; Larbot, A. Efficiency of grafting of Al_2O_3, TiO_2 and ZrO_2 powders by perfluoroalkylsilanes. *Colloid Surfaces A* **2013**, *420*, 64–73. [CrossRef]

42. Kujawski, W.; Kujawa, J.; Wierzbowska, E.; Cerneaux, S.; Bryjak, M.; Kujawski, J. Influence of hydrophobization conditions and ceramic membranes pore size on their properties in vacuum membrane distillation of water–organic solvent mixtures. *J. Membr. Sci.* **2016**, *499*, 442–451. [CrossRef]

43. Kujawska, A.; Kujawski, J.K.; Bryjak, M.; Cichosz, M.; Kujawski, W. Removal of volatile organic compounds from aqueous solutions applying thermally driven membrane processes. 2. Air gap membrane distillation. *J. Membr. Sci.* **2016**, *499*, 245–256. [CrossRef]

44. Kujawa, J.; Kujawski, W. Functionalization of ceramic metal oxide powders and ceramic membranes by perfluoroalkylsilanes and alkylsilanes possessing different reactive groups: Physicochemical and tribological properties. *ACS Appl. Mater. Interfaces* **2016**, *8*, 7509–7521. [CrossRef] [PubMed]

45. Kujawa, J.; Cerneaux, S.; Kujawski, W.; Bryjak, M.; Kujawski, J. How to functionalize ceramics by perfluoroalkylsilanes for membrane separation process? Properties and application of hydrophobized ceramic membranes. *ACS Appl. Mater. Interfaces* **2016**, *8*, 7564–7577. [CrossRef] [PubMed]

46. Efome, J.E.; Baghbanzadeh, M.; Rana, D.; Matsuura, T.; Lan, C.Q. Effects of superhydrophobic SiO_2 nanoparticles on the performance of PVDF flat sheet membranes for vacuum membrane distillation. *Desalination* **2015**, *373*, 47–57. [CrossRef]

47. Hosseinabadi, S.R.; Wyns, K.; Buekenhoudt, A.; Van der Bruggen, B.; Ormerod, D. Performance of grignard functionalized ceramic nanofiltration membranes. *Sep. Purif. Technol.* **2015**, *147*, 320–328. [CrossRef]
48. Rezaei Hosseinabadi, S.; Wyns, K.; Meynen, V.; Carleer, R.; Adriaensens, P.; Buekenhoudt, A.; Van der Bruggen, B. Organic solvent nanofiltration with grignard functionalised ceramic nanofiltration membranes. *J. Membr. Sci.* **2014**, *454*, 496–504. [CrossRef]
49. Fane, A.G.; Schofield, R.W.; Fell, C.J.D. The efficient use of energy in membrane distillation. *Desalination* **1987**, *64*, 231–243. [CrossRef]
50. Qtaishat, M.R.; Banat, F. Desalination by solar powered membrane distillation systems. *Desalination* **2013**, *308*, 186–197. [CrossRef]
51. Nene, S.; Kaur, S.; Sumod, K.; Joshi, B.; Raghavarao, K.S.M.S. Membrane distillation for the concentration of raw cane-sugar syrup and membrane clarified sugarcane juice. *Desalination* **2002**, *147*, 157–160. [CrossRef]
52. Sakai, K.; Koyano, T.; Muroi, T.; Tamura, M. Effects of temperature and concentration polarization on water vapour permeability for blood in membrane distillation. *Chem. Eng. J.* **1988**, *38*, B33–B39. [CrossRef]
53. Sakai, K.; Muroi, T.; Ozawa, K.; Takesawa, S.; Tamura, M.; Nakane, T. Extraction of solute-free water from blood by membrane distillation. *Am. Soc. Artif. Intern. Organ. Trans.* **1986**, *32*, 397–400. [CrossRef]
54. Kujawski, W.; Krajewska, S.; Kujawski, M.; Gazagnes, L.; Larbot, A.; Persin, M. Pervaporation properties of fluoroalkylsilane (FAS) grafted ceramic membranes. *Desalination* **2007**, *205*, 75–86. [CrossRef]
55. Kujawa, J.; Cerneaux, S.; Kujawski, W. Removal of hazardous volatile organic compounds from water by vacuum pervaporation with hydrophobic ceramic membranes. *J. Membr. Sci.* **2015**, *474*, 11–19. [CrossRef]
56. Kujawa, J.; Cerneaux, S.; Kujawski, W. Highly hydrophobic ceramic membranes applied to the removal of volatile organic compounds in pervaporation. *Chem. Eng. J.* **2015**, *260*, 43–54. [CrossRef]
57. Amirilargani, M.; Sadrzadeh, M.; Sudholter, E.J.R.; de Smet, L.C.P.M. Surface modification methods of organic solvent nanofiltration membranes. *Chem. Eng. J.* **2016**, *289*, 562–582. [CrossRef]
58. Owens, D.K.; Wendt, R.C. Estimation of the surface free energy of polymers. *J. Appl. Polym. Sci.* **1969**, *13*, 1741–1747. [CrossRef]
59. Thomas, R.; Guillen-Burrieza, E.; Arafat, H.A. Pore structure control of PVDF membranes using a 2-stage coagulation bath phase inversion process for application in membrane distillation (MD). *J. Membr. Sci.* **2014**, *452*, 470–480. [CrossRef]
60. Walsh, R. Bond dissociation energy values in silicon-containing compounds and some of their implications. *Acc. Chem. Res.* **1981**, *14*, 246–252. [CrossRef]
61. Choi, W.; Tuteja, A.; Mabry, J.M.; Cohen, R.E.; McKinley, G.H. A modified Cassie–Baxter relationship to explain contact angle hysteresis and anisotropy on non-wetting textured surfaces. *J. Colloid Interfaces Sci.* **2009**, *339*, 208–216. [CrossRef] [PubMed]
62. Woldemariam, D.; Martin, A.; Santarelli, M. Exergy analysis of air-gap membrane distillation systems for water purification applications. *Appl. Sci.* **2017**, *7*, 301. [CrossRef]
63. Ko, C.-C.; Chen, C.-H.; Chen, Y.-R.; Wu, Y.-H.; Lu, S.-C.; Hu, F.-C.; Li, C.-L.; Tung, K.-L. Increasing the performance of vacuum membrane distillation using micro-structured hydrophobic aluminum hollow fiber membranes. *Appl. Sci.* **2017**, *7*, 357. [CrossRef]
64. Kujawa, J.; Kujawski, W. Driving force and activation energy in air-gap membrane distillation process. *Chem. Pap.* **2015**, *69*, 1438–1444. [CrossRef]

MDPI AG

St. Alban-Anlage 66

4052 Basel, Switzerland

Tel. +41 61 683 77 34

Fax +41 61 302 89 18

http://www.mdpi.com

Applied Sciences Editorial Office

E-mail: applsci@mdpi.com

http://www.mdpi.com/journal/applsci

www.ingramcontent.com/pod-product-compliance
Lightning Source LLC
Chambersburg PA
CBHW051911210326
41597CB00033B/6108